AF362037

Advanced Techniques and Efficiency Assessment of Mechanical Processing

Advanced Techniques and Efficiency Assessment of Mechanical Processing

Editors

Daniel Saramak
Marek Pawełczyk
Tomasz Niedoba

MDPI • Basel • Beijing • Wuhan • Barcelona • Belgrade • Manchester • Tokyo • Cluj • Tianjin

Editors

Daniel Saramak
Department of Environmental
Engineering
AGH University of Science
and Technology
Krakow
Poland

Marek Pawełczyk
Department of Measurements
and Control Systems
Silesian University of
Technology
Gliwice
Poland

Tomasz Niedoba
Department of Environmental
Engineering
AGH University of Science
and Technology
Krakow
Poland

Editorial Office
MDPI
St. Alban-Anlage 66
4052 Basel, Switzerland

This is a reprint of articles from the Special Issue published online in the open access journal *Minerals* (ISSN 2075-163X) (available at: www.mdpi.com/journal/minerals/special_issues/ Mechanical_Processing).

For citation purposes, cite each article independently as indicated on the article page online and as indicated below:

LastName, A.A.; LastName, B.B.; LastName, C.C. Article Title. *Journal Name* **Year**, *Volume Number*, Page Range.

ISBN 978-3-0365-2875-5 (Hbk)
ISBN 978-3-0365-2874-8 (PDF)

Contents

About the Editors

Daniel Saramak

Expert in mineral processing, with over 20-years experience in the field gained at university, in domestic and international research centres and through international cooperation with industrial sector. Professor at AGH University of Science and Technology, Faculty of Mining and Geoengineering. Main research areas concern raw materials comminution, esspecially HPGR, optimization and modeling of selected operations of mineral processing, as well as economimc and technological assessment. Author and co-author of nearly 200 publications, coordinator and executor of many international and domestic research projects, co-author of international and domestic patents concerning raw materials processing technology. Extensive teaching experience related to raw materials, statistics, mathematics and economics.

Marek Pawełczyk

Scientific discipline (according to the Polish catalogue): Control, Electronic and Electrical Engineering; scientific specialties: industrial automation, digital signal processing, process identification, adaptive control, vibroacoustics, materials engineering, Industry 4.0.

Tomasz Niedoba

Assoc. Prof. Tomasz Niedoba is Professor at AGH University of Science and Technology in Krakow, Poland, representing Faculty of Civil Engineering and Resources Management, Department of Environmental Engineering. His specialization is statistical analysis and mathematical modeling. He concentrates on optimizing various processes, including mineral processing and environmental aspects. In his works, he tries to introduce new modern methods with an aspect of multidimensional analysis and visualization of such types of data. He is a tutor of Mathematics and Statistics and their applications in practice. Apart from science, he is interested in sport, literature and film.

Editorial

Special Issue "Advanced Techniques and Efficiency Assessment of Mechanical Processing"—Editorial Note and Critical Review of the Problems

Daniel Saramak 

Department of Environmental Engineering, Faculty of Civil Engineering and Resource Management, AGH University of Science and Technology, Mickiewicza 30 Av., 30-059 Cracow, Poland; dsaramak@agh.edu.pl

Citation: Saramak, D. Special Issue "Advanced Techniques and Efficiency Assessment of Mechanical Processing"—Editorial Note and Critical Review of the Problems. *Minerals* **2021**, *11*, 1428. https://doi.org/10.3390/min11121428

Received: 3 December 2021
Accepted: 16 December 2021
Published: 17 December 2021

Publisher's Note: MDPI stays neutral with regard to jurisdictional claims in published maps and institutional affiliations.

1. Introduction

The value chain of metal production consists of a number of processing steps that result in obtaining the final metal product from the given raw material. Each stage, from mining through to mineral processing and metallurgical treatment, is equally crucial and indispensable and performs a specific role in the entire mining and processing system. Mechanical processing is a step within the mineral processing stage. Its main duty is to provide material of a sufficient size, and thus generate an acceptable level of useful mineral liberation. This allows the required results to be achieved in downstream beneficiation operations, which are measured through the recovery of useful metals and the level of pay metal loss in tails. The primary technological operations associated with mechanical processing are comminution and classification. The aim of mechanical processing is, therefore, to provide a sufficient size reduction in material by dividing it into particles or fragments, which are measured using the comminution ratio level. All of these operations aim to remove impurities and other unwanted fractions from the feed, which is then mainly applied to the aggregate production sector. However, these operations are often treated as supplementary, since their incorporation into the technological circuit mostly depends on the qualitative characteristics and the contained impurities of the feed, which consists mostly of clayish and dusty fine fractions.

The mechanical processing stage is a relatively simple step and numerous investigations found in the literature recognised its technological solutions. However, despite its apparent simplicity, multistage crushing and grinding operations, including its separation and classification, make this processing stage more complicated than initially realised. The problem lies in the huge potential of its steering and control arrangement, as well as the variety of changeable operational parameters in the circuit. In light of this, three major groups of operational variables can be distinguished:

- Operational parameters of crushing and classifying devices;
- Physical and mechanical characteristics of the feed material;
- Technological regime and the manner of carrying out a specific operation.

The efficient operation of mechanical processing in technological circuits can be analysed from several points of view:

- Achieving the required technological outputs measured through the obtained size reduction in the feed material and specific particle size composition;
- Economical assessment of the real and potential effects through the analysis of the relationship between costs and benefits;
- Environmental and social aspects resulting from a general negative impact of mining on the environment and society;
- Application of optimization approaches and methods, together with simulation models based on theories of mathematical and statistical modelling;
- Other approaches not listed above or a combination of several scopes.

2. Methodology and Results

The assessment of the impact of a specific topic can be evaluated on the basis of a number of scientific publications concerning the issue. Bibliometric analysis is a good example of this approach, and by using this tool it is possible to perform qualitative and quantitative analyses of scientific publications registered in databases to obtain interesting results. In general, there are a relatively high number of publications concerning raw material treatment at the stage of mechanical treatment/processing but, to verify this, a detailed analysis must be performed, based on the records registered in the *Web of Science (WoS)* database. In the WoS *Core Collection*, the *all fields* option for searched documents was chosen, and the analysed period was *All years (1900–2021)*. The search was performed on 24 November 2021. The obtained search results (732,139 records) were then refined using the research area *Mining and mineral processing* in Web of Science Categories. As a result, 4285 records were obtained. The obtained records were analysed according to the year of publication, region/country, source of publication (journal name), research area, scientific institution, and the keywords. An analysis according to the keywords was performed separately for author words and for editor words. The CiteSpace application was used as a tool for obtaining the data and visualization. The number of publications registered in WoS database across individual years is presented in Figure 1.

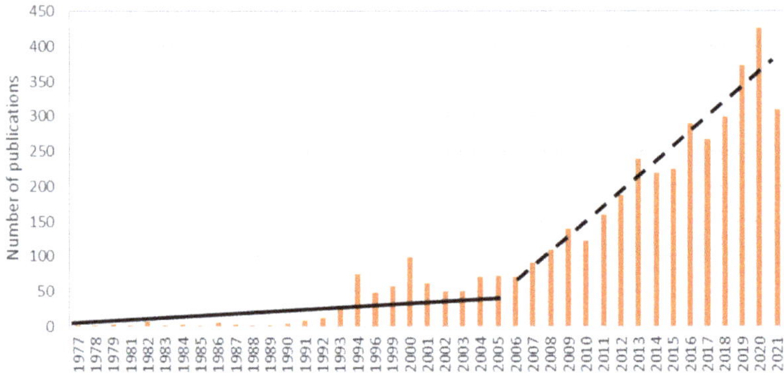

Figure 1. Number of publications issued yearly that concern mechanical processing of raw materials for the period 1977–2021.

As shown, two primary periods can be distinguished in a general trend:

- In the period between 1977 and 2005, the increase in the number of publications in consecutive years is very low; therefore, the trend can be determined as a constant (solid line in Figure 1);
- Since the year 2006, a regular increase in publication numbers can be observed for each consecutive calendar year. This trend can be approximated using a linear function (dashed line in Figure 1).

The above results demonstrate that problems concerning the mechanical processing of raw material generated significant interest, especially in the last decade. Next, an analysis of publications regarding regions and countries is presented. The number of publications according to specific countries are presented in Table 1, where the top 10 countries with the most publications are shown. The results show that nearly 30% of the world's publications regarding mechanical processing technology of raw materials originate from China. The USA is the second country with the most publications, totalling around 17%, with 741 papers. Australia is the next country, with its share of publications exceeding 7%, closely followed by Germany and Russia, who produced 254 and 237 publications, respectively.

Table 1. Top 10 countries with the highest number of publications in Web of Science database.

Country	Number of Publications	Percentage Share
China	1226	28.638
USA	741	17.309
Australia	313	7.311
Germany	254	5.933
Russia	237	5.536
Canada	203	4.742
Poland	200	4.672
India	164	3.831
Iran	156	3.644
England	113	2.640

Publications grouped according to research areas are presented in Table 2. All records are within the main research area, i.e., *Mining and mineral processing*, but most publications have more than one assigned research area. The three most popular areas with more than one thousand publications are *Metallurgy, Metallurgical Engineering, Mineralogy* and *Materials Science Multidisciplinary*. It is worth mentioning that a single record (publication) can be assigned to several areas; therefore, the cumulative percentage share in Table 2 is higher than 100%. This is clearly visible based on the three mentioned categories. The reason for this is that the cumulative percentage share for them only exceeds 100 percent; many publications concerning the problems of mechanical processing must be associated with at least two of the top three areas.

Table 2. Most frequent research areas associated with the publications concerning the issue.

Research Area	Number of Publications	Percentage Share
Metallurgy Metallurgical Engineering	1739	40.583
Mineralogy	1631	38.063
Materials Science Multidisciplinary	1567	36.569
Engineering Geological	575	13.419
Engineering Chemical	534	12.462
Geosciences Multidisciplinary	220	5.134
Geochemistry Geophysics	203	4.737
Energy Fuels	75	1.750
Chemistry Physical	65	1.517
Environmental Sciences	63	1.470

An analysis of obtained records from the WoS database, grouped according to journals, is presented in Table 3, including the top 20 journals with the highest number of publications related to mechanical processing. The top journal on this list, *JOM*, has an almost 20% share of the number of published articles, and a significant gap can be observed between the first and second position on the list.

An analysis of the *impact factor* value shows that the number of publications in a specific journal is, to some extent, correlated with IF. However, this link is not very strong and is influenced by several factors, mostly by the assignation of individual titles to different disciplines and categories. If there is an asterisk at the value of an individual IF, it denotes that this value is older than the year 2020. Table 3 also presents the positions of a specific journal determined by the number of quartiles in the *Mining and mineral processing* category. From this, two conclusions can be drawn. First, detailed profiles of journals differ and second, not all the journals are equally interested in problems concerning the mechanical aspects of mineral processing. In turn, the number of publications according to individual scientific institutions is presented in Table 4. The results show that among the five topmost publishing institutions, three are located in China, with almost a 12% share.

Table 3. Top 20 journals with the highest number of publications concerning mechanical processing.
*: the value of an individual IF; -: this value is older than the year 2020.

Title of Journal	Number of Publications	Percentage Share	IF	Quartile
JOM	798	18.641	2.474	2
International Journal of Rock Mechanics and Mining Sciences	398	9.297	7.135	1
International Journal of Minerals Metallurgy and Materials	319	7.452	2.232	3
Minerals Engineering	289	6.751	4.765	1
International Journal of Mineral Processing	229	5.349	2.688 *	-
Minerals	133	3.107	2.644	2
JOM Journal of The Minerals Metals Materials Society	115	2.686	-	-
Journal of University of Science and Technology Beijing	106	2.476	0.919 *	-
International Journal of Mining Science and Technology	76	1.775	4.084	1
Archives of Mining Sciences	64	1.495	1.127	4
Physicochemical Problems of Mineral Processing	64	1.495	1.213	4
Advanced Materials Research	60	1.402	-	-
Journal of Mining Institute	57	1.331	-	-
Minerals Metals Materials Series	55	1.285	-	-
Mining of Mineral Deposits	55	1.285	-	-
Acta Geodynamica et Geomaterialia	49	1.145	1.176	4
Acta Montanistica Slovaca	48	1.121	1.413	3
Journal of The South. African Institute of Mining and Metallurgy	48	1.121	0.807	4
Journal of Mining Science	44	1.028	0.456	4
Mineral Processing and Extractive Metallurgy Review	44	1.028	5.283	1

Table 4. Top 10 publishing institutes.

Research Area	Number of Publications	Percentage Share
University of Science Technology Beijing	228	5.326
China University of Mining Technology	184	4.298
United States Department of Energy Doe	129	3.013
Russian Academy of Sciences	94	2.196
Chinese Academy of Sciences	93	2.172
Northeastern University China	88	2.056
Central South University	79	1.845
University of Queensland	69	1.612
Helmholtz Association	68	1.588
AGH University of Science Technology	60	1.402

3. Qualitative Analysis of the Content and Discussion

The analysis below concerns the quality of content in articles connected with the mechanical processing of raw materials. The analysis was performed both on the basis of keywords included in the articles that were connected with mechanical processing and indexed in the WoS database, and through a content review of the papers published within this Special Issue. Two types of keywords were analysed: author keywords, and editorial keywords, i.e., phrases added by the editorial boards of the journal.

3.1. Keywords Analysis

An analysis of author keywords indicated that the top ten most used keywords constituted almost one-third of records (Table 5). It can also be seen that a majority of these keywords were associated with environmental aspects.

Table 5. Top 10 author and editor keywords, common phrases in bold font.

Author Keywords	Percentage Share	Number of Cases	Editor Keywords	Percentage Share	Number of Cases
air pollution	11.67	3088	**particulate matter**	2.93	3336
particulate matter	10.01	2648	pm10	2.91	3312
air quality	3.75	993	**air pollution**	2.66	3036
source apportionment	1.49	394	pm2.5	2.44	2780
air pollutant	1.41	373	pollution	2.03	2316
indoor air quality	1.22	322	particle	1.87	2129
heavy metal	0.98	258	exposure	1.82	2069
nitrogen dioxide	0.70	184	aerosol	1.61	1837
chemical composition	0.70	184	mortality	1.53	1748
size distribution	0.60	160	emission	1.30	1480

Two phrases are predominant regarding the number of cases; their share exceeds 20%, and both of them are connected with environmental problems. However, this only confirms a global trend observed in the mineral processing sector, which aims to pay more attention to environmental aspects. Phrases strictly typical to mineral processing are also on the top 10 list, but they are located at the end of the list. An analysis of editor keywords indicated the likelihood that, in the case of the author keywords, the share of individual phrases was more equally distributed, and it was not possible to clearly distinguish a single predominant word. However, similar to the author keywords, editor keywords were mostly related to environmental problems, only confirming that the greatest concern of the raw material sector is its impact on the environment. More detailed visualizations are shown in Figures 2 and 3.

3.2. Content Analysis

An analysis of SI content suggested that the dominating topic was technology. A detailed analysis of the content shows that the majority of papers concern various methods of classification and separation: from screening classification [1,2] and sorting [3], to jig [4] and flotational [5] beneficiation, with one publication related to crushing [6]. Authors applied various methods to monitor the obtained qualitative characteristics [1,3] and utilized simulation and modelling tools [2,5,7] in order to evaluate the operation effectiveness of individual separation techniques. Technological aspects were also underlined in a review of recent directions of comminution technology development [8]. It is accepted that the mechanical processing stage predetermines the efficiency of downstream separation operations, especially in the technological circuits of ore beneficiation. For industries associated with the processing of rock materials, mechanical processing, in turn, directly influences the characteristics of final products in terms of size and shape [9]. This was the subject of investigations in papers by Saramak, and Gawenda et al. [4,8].

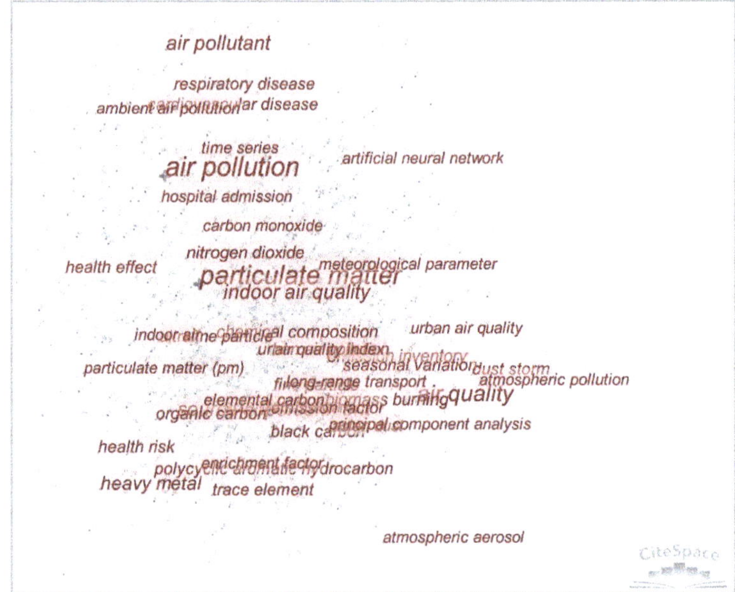

Figure 2. Visualization of author keywords.

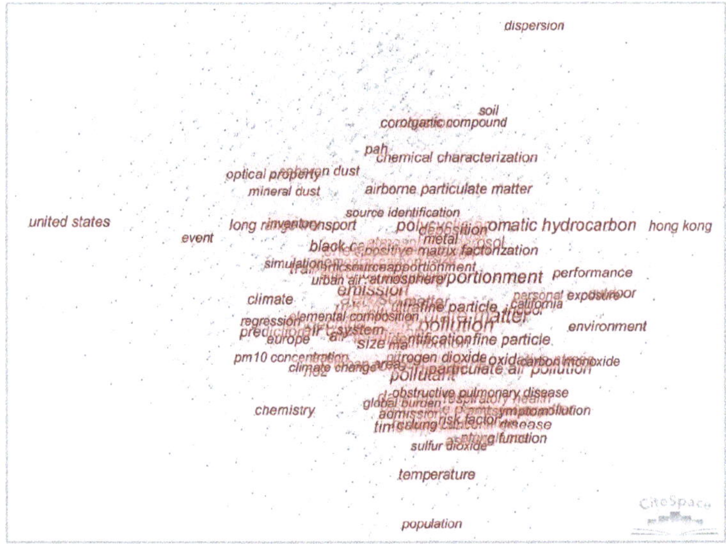

Figure 3. Visualization of editor keywords.

Screening technology was the scope in works of N. Duda-Mróz et. al. And C. Yu et. al. [1,2]. In the work [1] focuses on development of diagnostic procedures and monitoring of screening device, particularly its spring, to estimate the time needed for the safe operation of the device and early damage detection. One approach consisted of an analysis of vibration data registered for the specific components of the device, i.e., springs and bearings, and identification of disturbances by means of calculated techniques of wavelet filtering. The second paper, concerning size classification [2], focuses on the screening of high-viscosity, fine, cohesive particles. The authors analysed the achieved screening efficiency from the scope of the device operation (different sections of the mat surface) and surface energy of particles. The authors also utilized the DEM modelling technique to simulate the behaviour of material on the screen. They also concluded that the service energy level of the particles regarding adhesion was inversely proportional to the screening efficiency. The study of [3] concerns the sorting technology and investigates the system of Dual-Energy X-ray (DE-XRT) and its ability to distinguish sulphides from non-sulphides. For this type of material, it was possible to achieve a very high accuracy and the system appeared to be an effective sensor that could be used to differentiate sulphides from waste material. The issues presented in this work have an advantage over recent achievements in the development of visual methods and the increasing sensitivity and accuracy of detecting tools [10]. These methods are significant from the scope of the proper characterization of granular material, especially in its description according to particle size, but also for other features, such as the density, porosity, or content of different types of materials. These issues can be found in paper [7], where problems concerning granular material segmentation during transportation are analysed. The practical aspects of this work seem to be significant due to the possibility of online monitoring and the early detection of improper or undesired states. The application of neural networks algorithms makes the system more effective. This is especially significant from the scope of material characterization and helps to select and optimize the downstream separation techniques of granular material. Similar aspects were included in the work of [1]. Optimizational models were also developed in the work of Niedoba et al. [5], which applied copper ore flotation. A mathematical model based on taxonomic methodology was developed, as well as the adopted functions enabling the determination of the optimal technological parameters of flotation, depending on the material characteristics and process regime. This approach is in line with the popular direction of modelling in mineral processing; assuming the binding of parameters in theoretical models or functions with a material, device, and process course [2,8,9,11].

Rock materials enrichment is also present in studies concerning the beneficiation of aggregates in a jig device [4]. This problem is common for a wide group of rock materials, confirming the opinion expressed in other papers of this Special Issue: that the proper characterization of granular material is of key significance in the effective separation of rock materials. The utilization of a patented system of material classification, upstream to the beneficiation, significantly improves separation according to size and shape, but the selection of operational parameters for this system is possible due to the knowledge on the size and shape characteristics of the processed material.

One paper is strictly related to comminution [7], concerning the problem of material breakage in dynamic conditions. Due to the complexity of this process, resulting from a series of interactions occurring both in space and time, this problem is under constant investigation. This paper's findings are interesting as they concern the general problem of raw materials breakage, namely the low effective utilization of energy for breakage and high energy-consumption. The results can contribute to a better understanding of selected mechanisms relating to dynamic breakage, such as the probabilities of establishing the given size of particles and their location within the crushed product.

4. Conclusions

Papers included in this Special Issue of *Minerals*, entitled "Advanced Techniques and Efficiency Assessment of Mechanical Processing" addressed significant problems related to the processing of raw materials that can be grouped into the following categories:

- Methods and techniques of monitoring and the visual characterization of granular materials;
- Modelling and optimization of process effectiveness, using advanced computational tools and algorithms;
- Efficiency assessment of selected operations that are crucial in the mineral processing industry, and can be performed from a technological or economic point of view.

Several aspects of this Special Issue were found to be related to environmental problems, especially in papers concerning the treatment of rock materials. An analysis of keywords in articles registered in the WoS database indicated that issues concerning the environment are of key significance. Its role in scientific research is therefore becoming increasingly important.

Acknowledgments: The Guest Editor thanks all authors of articles included in this Special Issue, as well as reviewers, Assistant Editors and the Editorial Board for their input and work towards this Special Issue. Special thanks also due to the Bibliometric Analysis Group from the Main Library of AGH University of Science and Technology for support and valuable comments on the bibliometric data analysis.

Conflicts of Interest: The author declare no conflict of interest.

References

1. Duda-Mróz, N.; Anufriiev, S.; Stefaniak, P. Application of Wavelet Filtering to Vibrational Signals from the Mining Screen for Spring Condition Monitoring. *Minerals* **2021**, *11*, 1076. [CrossRef]
2. Yu, C.; Geng, R.; Wang, X. A Numerical Study of Separation Performance of Vibrating Flip-Flow Screens for Cohesive Particles. *Minerals* **2021**, *11*, 631. [CrossRef]
3. Zhang, Y.; Yoon, N.; Holuszko, M.E. Assessment of Sortability Using a Dual-Energy X-ray Transmission System for Studied Sulphide Ore. *Minerals* **2021**, *11*, 490. [CrossRef]
4. Gawenda, T.; Saramak, D.; Stempkowska, A.; Naziemiec, Z. Assessment of Selected Characteristics of Enrichment Products for Regular and Irregular Aggregates Beneficiation in Pulsating Jig. *Minerals* **2021**, *11*, 777. [CrossRef]
5. Niedoba, T.; Pięta, P.; Surowiak, A.; Şahbaz, O. Multidimensional Optimization of the Copper Flotation in a Jameson Cell by Means of Taxonomic Methods. *Minerals* **2021**, *11*, 385. [CrossRef]
6. Guo, Q.; Pan, Y.; Zhou, Q.; Zhang, C.; Bi, Y. Kinetic Energy Calculation in Granite Particles Comminution Considering Movement Characteristics and Spatial Distribution. *Minerals* **2021**, *11*, 217. [CrossRef]
7. Ma, X.; Zhang, P.; Man, X.; Ou, L. A New Belt Ore Image Segmentation Method Based on the Convolutional Neural Network and the Image-Processing Technology. *Minerals* **2020**, *10*, 1115. [CrossRef]
8. Saramak, D. Challenges in Raw Material Treatment at the Mechanical Processing Stage. *Minerals* **2021**, *11*, 940. [CrossRef]
9. Ambróst, W. Jigging: A review of fundamentals and future directions. *Minerals* **2020**, *10*, 998. [CrossRef]
10. Gawenda, T.; Krawczykowski, D.; Krawczykowska, A.; Saramak, A.; Nad, A. Application of Dynamic Analysis Methods into Assessment of Geometric Properties of Chalcedonite Aggregates Obtained by Means of Gravitational Upgrading Operations. *Minerals* **2020**, *10*, 180. [CrossRef]
11. Barrios, G.K.P.; Tavares, L.M. A preliminary model of high pressure roll grinding using the discrete element method and multi-body dynamics coupling. *Int. Journ. Min. Proc.* **2016**, *156*, 32–42. [CrossRef]

Review

Challenges in Raw Material Treatment at the Mechanical Processing Stage

Daniel Saramak 

Department of Environmental Engineering, Faculty of Civil Engineering and Resource Management, AGH University of Science and Technology, 30-059 Cracow, Poland; dsaramak@agh.edu.pl

Abstract: This paper concerns problems related to the mechanical processing of mineral raw materials. The aspects explored were limited to the analysis of comminution technologies in terms of their effectiveness and energy consumption, modeling and simulation approaches, the assessment of crushing results, and environmental aspects. This article includes investigation of new technologies of comminution, comparing HPGR, high-voltage pulses, and electromagnetic mills. In the area of modeling and optimization, special attention was paid to the approximation of the particle size distribution of crushing products by means of Weibull, log-normal, and logistic functions. Crushing products with an increased content of fines were well characterized by Weibull's distribution, while log-normal function adequately described HPGR products with a relatively low content of fines.

Keywords: raw materials; mineral processing; enrichment; comminution; HPGR; approximation of particle size

Citation: Saramak, D. Challenges in Raw Material Treatment at the Mechanical Processing Stage. *Minerals* **2021**, *11*, 940. https://doi.org/10.3390/min11090940

Academic Editor: Thomas Mütze

Received: 24 July 2021
Accepted: 25 August 2021
Published: 29 August 2021

Publisher's Note: MDPI stays neutral with regard to jurisdictional claims in published maps and institutional affiliations.

1. Introduction

The processing technology used for mineral raw materials is crucial in terms of the production of numerous metals and non-metallic products sourced and extracted by means of mining techniques. Though the entire value chain of metal production includes a number of steps (from geology, through to mining, metallurgy, and manufacturing), the mineral processing stage, to some extent, influences the quality of the final product and the effectiveness of the entire process of commercial product manufacturing. Despite its complexity and the application of many physical, mechanical, chemical, and other types of separation processes and operations, ore mineral processing can generally be divided into two overarching stages:

(a)   reduction in the size of the feed material;
(b)   separation of useful mineral from the gangue.

The primary purpose of the size reduction stage is the liberation of useful mineral in a way that allows for proper separation in the downstream separation stage. This process for the case of ores is presented in Figure 1. The ROM (run-of-mine) material is a mix of compounds of ferrous or non-ferrous metals and gangue. Useful elements and compounds (black color on Figure 1) are "locked" among the gangue material in the ROM. In such a form, it is rather difficult to separate them from the gangue with high efficiency. Through the application of comminution operations, the size of individual particles can be made finer and thus the liberation of useful compound is more intense compared to the ROM material (center of Figure 1). Compared to non-crushed ores, the separation of such materials can be carried out in a less complicated manner and higher recoveries can be achieved. The separation product contains a significant amount of useful compounds (lower part of Figure 1). This situation applies to ore enrichment, which are generally divided into ferrous and non-ferrous minerals (i.e., cupriferous, gold or silver-bearing, and other metals except for iron).

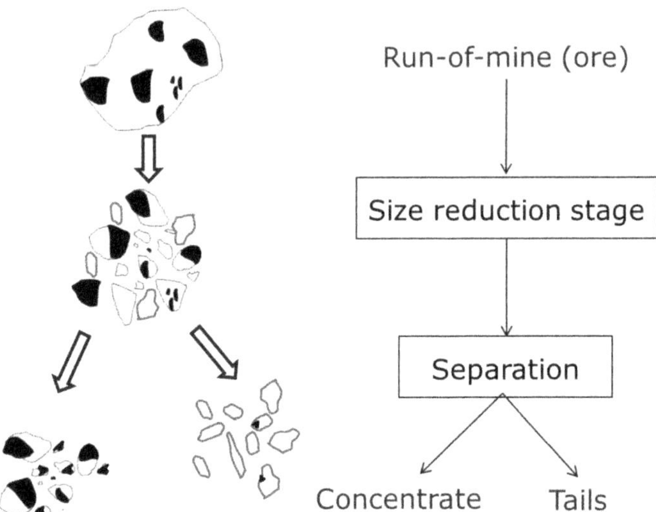

Figure 1. Mineral processing for metal-bearing ores.

In both aggregates production and the processing of rock materials, (such as crushing stones, sands, and gravels), the stage of size reduction (comminution) constitutes the primary step. This directly influences the qualitative parameters of final products in terms of size and shape (aggregate production sector) or specific surface (cement clinker industry) [1].

Considering previous text, the mechanical processing of raw materials is a crucial stage in the technology of both ore processing and the aggregate production sector. Proper passage through this stage initially influences the potential effectiveness of the core beneficiation operations of minerals from the scope of useful mineral recovery, and much research has been carried out in this area. Current trends in these investigations mainly include:

- development of new comminution technologies;
- optimizing the performance of recent applications;
- modeling, simulation, and performance optimization of mechanical processing circuits.

Recently, environmental aspects have been gaining greater attention, especially in terms of decreasing the negative impact of mechanical processing operations on the environment and society; in particular, attention has been paid to harmful and annoying emissions of dusts, gases, heat, and noise. Operations of industrial comminution account for as much as 3% to 5% of the total usage of electric energy in the world [2]. Despite this, their performance effectiveness is relatively low and has an inverse relationship to the size of the crushed material. For example, the energetic efficiency of tumble mill operation is estimated to be around 15%. Meanwhile, the production efficiency of some types of very fine grinding products can be as low as 1% [3]. Comminution energy utilization is higher in crushers, particularly in impact crushers. However, it is dependent on the type of raw material used and the crushing stage within the technological circuit [4]. Research generally confirms that crushing uses between 0.5 and 1 kWh/Mg, and that crushing devices using attrition and shear forces consume between 40% and 80% more comminution energy than machines based on impact forces [5]. Compression and impact forces techniques in material disintegration save the most energy. However, the low energy utilization for a breakage mechanism in conventional crushing and grinding equipment leads to the development of innovative comminution technologies [6].

2. Development in Crushing and Grinding Technology

2.1. HPGR Technology

HPGR devices are considered one of the most energy-efficient comminution machines. However, despite their lengthy existence in industrial mineral processing circuits, there have been many research projects aimed at improving their operational performance. While HPGRs indisputable energetic benefits have been largely proven in the literature [7,8], results from more recent investigations show room for improvement in both separation in flotation and leaching processes [9–11], mostly due to the intense liberation of useful mineral [12].

The average unit energy consumption for HPGR devices varies between 2 and 3–4 kWh/Mg; however, this depends on the mechanical properties of the material, the equipment size, and the recirculation scheme of the HPGR feed and product. Industrial practice shows that in calculations the value of 2.5 kWh/Mg is quite often accepted. Compared to tertiary crushing devices, applying HPGR to the comminution circuits of ore lowers grinding energy by 20–30% in downstream operations.

In the SAG-based grinding circuit, the ball mill grinding energy consumption is comparable to the grinding energy used in mills following the HPGR. However, the overall energy reduction in a HPGR-mill configuration can be greater than 30%. The benefits of HPGR in mineral beneficiation are visible at the level of useful mineral liberation. Microscopic analysis has found [12] values of 75 to 95% for copper minerals liberated from sulfide ores, depending on the operating pressure value. Copper recovery in downstream flotation varies from 80 to 85%, while the same value for ore that has been conventionally crushed (i.e. without using of high-pressure technology) is 80%. HPGR is beneficial in the recovery of minerals, as well as in leaching and cyaniding operations. Copper extraction from sulfide ores was 2 to 8% higher when leaching from HPGR-based circuit products, and kinetic readings of the process were more favorable [13]. Comparable effects are evident when heap leaching other minerals from upstream ores treated by HPGR. The metal extraction rate for HPGR is 10–15% higher than that of conventional crushing devices [14].

2.2. High Voltage Breakage

Investigations into improving the efficiency of breakage energy led to the development of other innovative techniques of comminution. High-voltage electrical pulse technology helps to achieve more intense liberation in ore comminution. The efficiency of this technique greatly depends on the regime of pulsation, as well as the texture and mechanical and electrical properties of the ore material. The optimization of pulse parameters is considered a very significant issue; however, it also creates problems, and a pulse generator dedicated especially to mineral liberation should be used [15]. Laboratory practice shows that if electric pulses are not adjusted to the material properties, there is no improvement in the effectiveness of liberation compared to using mechanical comminution techniques. In one example, gold ore which had been treated upstream by electric pulses did not record significantly improved extraction compared to mechanical breakage. However, after the optimization of the pulsating regime, the gold recovery was 35% higher due to the more intense liberation [16].

Gold recovery in cyanidation can be 15–50% higher than in that in ores that have been mechanically ground; however, this depends on the material's properties. In some cases, a 60–70% recovery can be observed.

Lower rates of metal recovery in flotation can be found; however, some results show that up to a 20% higher concentration of metal can be obtained in tail flotation following high-voltage pulses. Estimates of the unit energy consumption for the high-voltage pulse technique show that it varies from 3 to 5 kWh/Mg. There are relatively few publications in world scientific databases concerning investigations on the using of high-voltage pulses in comminution. The bibliometric analysis of the Core collection of Web Science database searching for the terms "high voltage pulses" and "comminution" or "liberation" shows that 100 publications have been published on this topic since 2012.

2.3. Electromagnetic Mills

The grinding technology used in electromagnetic mills is a novel method of material disintegration that utilizes rotating electromagnetic fields [17]. Fast-moving grinding media in the shape of short rods, made of ferromagnetic material, cause material breakage in the working chamber. Laboratory tests confirm the very short duration of the process; it requires from several to few tenths of seconds to prepare a 500 g sample for flotation operation. In terms of comminution ratio, the achieved results are very good compared to those achieved using conventional tumbling mills, thus the level of liberation is also higher. The technology is relatively new; only 50 publications can be found in international scientific databases, with nearly 75% issued within the last 5 years. Similarly to high-voltage pulses, the main issue is the lack of full-scale plant installation. However, in this case a quarter-scale machine is available. No test results for ore beneficiation in hydrometallurgy operations for electromagnetic mill products have been obtained so far; however, the results of flotational separation for sulfide copper ore show copper recovery increased by about 10–15% compared to conventional crushing and grinding devices. It is, however, difficult to find indisputable evidence for overall separation improvement, especially at the semi-plant and plant scale; however, it seems that for some specific purposes, the technology may bring be beneficial.

Table 1 summarizes the three characterized technologies for raw material comminution in terms of their capacity, energy consumption, and technological effectiveness.

Table 1. Summary of the benefits of HPGR, electric pulse breakage, and electromagnetic mills.

Type of Benefit	Unit	Type of Comminution Technology		
		HPGR	Electromagnetic Mill	Electric Pulses
Unit energy consumption	kWh/Mg	2–4	50–150	3–5
Benefits in mineral liberation compared to conventional crushing	%	10–20	10–15	10–15
Benefits in useful mineral recovery compared to conventional crushing: hydrometallurgy	%	2–8	No data	15–70
Benefits in useful mineral recovery compared to conventional crushing: flotation	%	1–4	5–20	up to 20
Plant-scale operation of the technology	-	Yes	No	No
Energy savings compared to conventional crushing circuit	%	Up to 30	No data	No data
Capacity increases compared to conventional crushing circuit	%	0–15	No data	No data

3. Circuits Layout and Optimization

3.1. Design Assumptions

A suitable circuit layout, with properly designed material flows among specific operations, including recycle streams, is of great significance in optimizing feed disintegration processes. Following the one overarching principle of mechanical processing: "do not crush unnecessarily", special attention is (or, at least, should be) paid to the by-passing of fully or sufficiently liberated particles [18]. This is especially true for comminution techniques that cause greater generation of micro-cracks during grain disintegration processes which may contribute to higher liberation. Such materials are proceeded to pre-concentration, where some share of useful minerals can be quickly recovered. Such an approach can be characterized by the following effects:

- A reduction in the capacity requirements of downstream grinding stage(s), which allows the installation of smaller grinding devices. More intense disintegration occurs in upstream crushing processes, which requires much less energy than grinding operations (Figure 2). The Figure gives a general idea of relationship between the size of treated material and the required energy for comminution. An exponential increase in the grinding energy can be observed, together with further decreasing the size of already fine material;

- The optional application of separation within comminution circuits, especially devices based on physical separation, such as jigs [19]. Jig separation has a long history of existence in technological circuits of mechanical processing or raw materials; however, many investigations aiming to improve the process for specific conditions and individual materials have been carried out [20,21]. This cost-efficient technology may give favorable results in the extraction of useful mineral amongst certain particle size fractions. Available results also confirm the potential for the separation of materials with relatively low differences in densities, provided specially designed devices are used [22].

- Decrease in the grinding energy consumption, which decreases the overall energy consumption of the circuit operational costs.

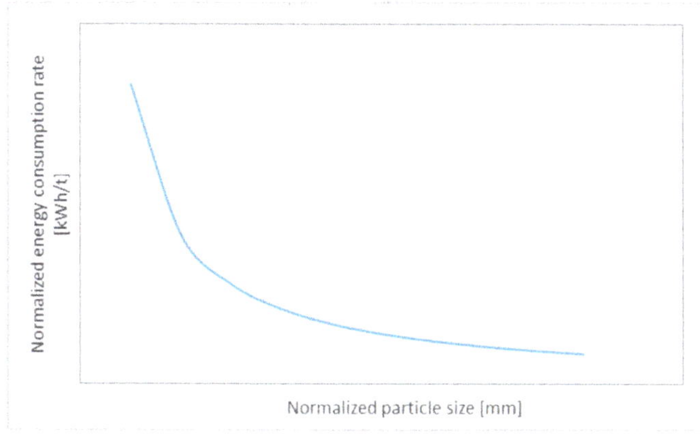

Figure 2. Relationship between energy usage for breaking and the size of treated particles.

Economic benefits include a reduction in CAPEX due to the possibility of the installation of smaller size devices, as well as contributing to the reduction in a major component of OPEX.

Such non-conventional layouts—i.e., where the breakage is more intensive on earlier comminution stages—are popular, especially in HPGR-based circuits. The relative energy savings in such configurations may reach 25% [23].

3.2. Simulation in Mineral Processing

Nowadays, many techniques and methods supporting the design, efficient operation, and performance assessment of mechanical processing are used in investigations. Among the most significant characteristics determining the effectiveness of comminution operations are particle breakage intensity and size reduction ratio, the size distribution of products, productivity, energy consumption, wear of liners, and movable parts of machines and others [24,25]. These can be distinguished following major directions in the modeling of mineral processing:

- Simulation tools and techniques showing models of behavior of grained material during the specific process, operation of a device, interactions amongst particles of the material, and interactions between the material and device. Various numerical techniques can be used in these simulations. The Discrete Element Method (DEM) is an especially popular technique used for this process, as well as in other disciplines outside mineral processing.

- Models capable of predicting the specific results of process performance. These include the approximation of particle size of comminution product, screening efficiency, comminution ratio, volume of material recycle, and others. These models are either based on theoretical distributions of random variables with confirmed applications in mineral processing or utilize principles of mathematical modeling.
- Optimization tools and applications based on the principles of mathematical and statistical modeling [26–28].

Modeling and simulation techniques for mapping the behavior of discrete mediums, such as grained materials, have become useful and powerful tools as the computational power and potentials of dedicated hardware accelerated in the last few decades. As a result, it is now possible to carry out very large numbers of calculations in very short time sequences for each modeled element. It is also now possible to characterize short-distance interactions (i.e., particle–particle, particle–device) as well as long-range impacts, (i.e., gravity, electrostatic, magnetic, or other forces). The mentioned DEM method gained significant popularity in the modeling of various enrichment processes, and many examples concerning this issue are present in the literature.

The mineral processing constitutes only a fraction of the wide practical usage of this simulation technique, but some significant aspects of mineral processing were covered to some extent:

- Description of granular material flow in comminution processes;
- Potential prediction of particle size distribution for selected crushing products;
- Equipment design on the basis of analyzed process behavior;
- Simulation of conveyor transportation operations;
- Description of motion of particles in selected processes of gravitational separation.

There are, however, gaps and challenges for this method, especially in the simulation of fine and very fine particle motion, (i.e., in fine grinding processes). The problem is that a number of particles in simulation is limited, what requires the use of a method of extrapolation, especially for finer sizes.

4. Modeling Approach

The use of the correct modeling approach assists in the description of an operation, as well as the results achieved for specific operations in mineral processing. These have significant cognitive meaning, especially for conditions beyond the operational regime, such as an increased throughput, exceeded values of operational parameters of devices (i.e., increased F_{sp} in HPGR, higher rotational speed of shaft in impactors, higher/lower amplitude/frequency of vibration screen), and others. These situations, however, are undesirable and efforts have been made to eliminate them or at least to limit their impact on the process course. The issue of greater significance is the possibility of the assessment results of mechanical processing proceeded through specific operation. Comminution results seem to be the most important among them, and the most popular approach consists in approximation of PSD of crushing products [29]. The second significant issue seems to be assessment of the useful mineral liberation degree, but nowadays it appears that only analytical methods utilizing scanning electron microscopy (SEM), such as MLA, can be effective. The PSD approximation method is not new and has been present in mineral processing investigations for decades. It consists of the assumption that the act of particle breakage is a kind of probability, especially concerning the size of newly created particles in the crushing product [30,31]. The approximation of crushing results technically utilize the Least Squares method (LS) and can be performed by means of various mathematical functions (distributions). Most of these have a confirmed application into a specific type and size of feed material and crushing device [32]. For example, the Weibull (or RRB) distribution was introduced into mineral processing in the 1930s [33] for the assessment of the particle size of crushing products. Log-norm distribution, in turn, has found application for the approximation of fine crushing (grinding) products [34]. In recent decades, functions previously used in other disciplines have also been used in mineral processing. The logistic distribution could be such an example. Approximation functions used in the estimation of particle size distribution of crushing products usually have two parameters, denoted as shape and scale parameters. It is worth mentioning that the application of theoretical distributions with a greater number of parameters may improve the modeling results; however, problems can appear in the interpretation of results, especially when the model is used in more general cases. So-called "censored distributions", (i.e., theoretical distributions with additional parameters that limits the particle size of the product by introducing minimum or maximum particle (d_{min} or d_{max})) can be more efficient in some cases [35].

The fitting accuracy can be assessed through the estimation error s_r, which is defined through Formula (1).

$$s_r = \sqrt{\frac{\sum_{i=1}^{n}\left(y_{i_{emp}} - y_{i_{mod}}\right)^2}{n-2}}, \tag{1}$$

where $y_{i_{emp}}$, empirical data; $y_{i_{mod}}$, modeling data; n, number of data points.

An exemplary approximation of the particle size distribution of a HPGR product was performed using three approximation functions, as presented in Table 2.

Table 2. Approximation functions used in fitting and the obtained results.

Approximation Function	Approximation Formula
Weibull's distribution	$F(d) = 1 - \exp\left(-\frac{d}{d_0}\right)^n$
Log-norm distribution	$F(d) = \frac{1}{2\pi}\int_{-\infty}^{t}\exp\left(-\frac{t^2}{2}\right)dt, \quad t = \frac{\ln\left(\frac{d}{d_0}\right)}{\sigma}$
Logistic distribution	$F(d) = \frac{1}{1+b\cdot exp(-c\times d)}$

The testing program included the crushing of feed material in the laboratory HPGR device under two values of operational pressing forces F: 15 and 10 kN. The width of rolls L = 100 mm, diameter D = 300 mm. Three samples of the same feed material with a similar particle size distribution were crushed in the HPGR press device under various conditions in order to obtain products with diverse particle size distributions:

- product 1: the material with increased content of fines, the feed was crushed twice under base pressing force (15 kN);
- product 2: the material with relatively lower content of fines—crushed under lower pressing force (10 kN);
- product 3: the material with balanced content of individual particle size fractions—crushed once under base pressing force (15 kN).

The PSD of all products, along with fittings with three functions, are presented in Figures 3–5.

The test results clearly show that the results of comminution depend on the manner of crushing device operation. However, it is also evident that the results of the comminution in each case can be approximated with different effects depending on the function used in calculations. It appears evident that crushing products with an increased content of fines can be described well through the use of Weibull's formula. The HPGR crushing product obtained for average crushing force turned out to be very well characterized by logistic distribution.

When the feed material was crushed under a lower pressing force, the particle size composition of the obtained product could be adequately described with the use of a log-normal distribution. Detailed characteristics of the approximation in terms of fitting error values are given in Table 3.

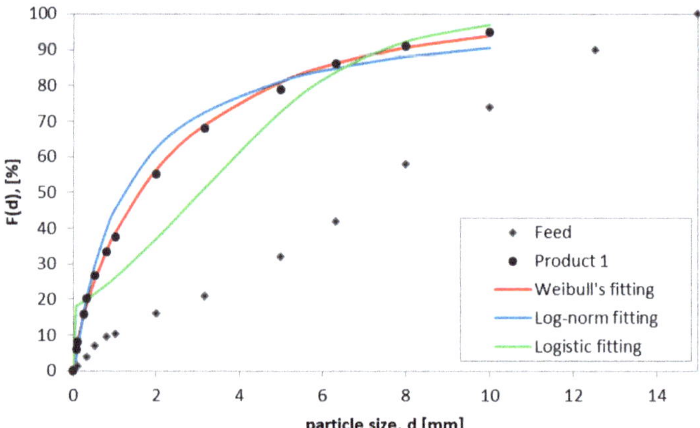

Figure 3. Fitting results for product 1.

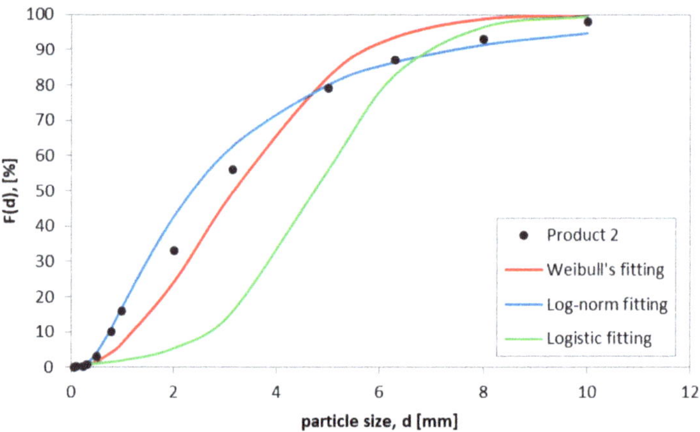

Figure 4. Fitting results for product 2.

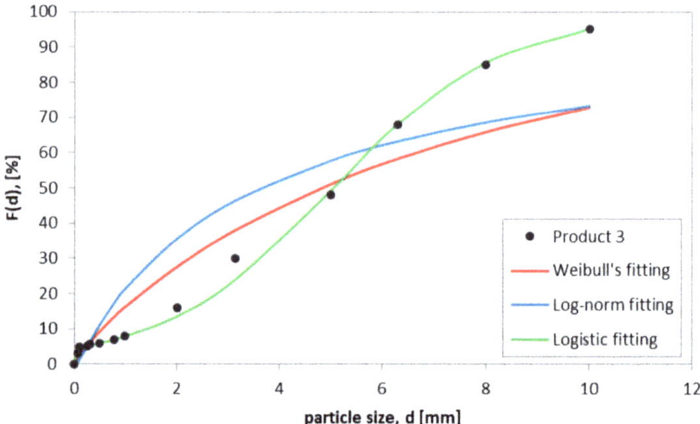

Figure 5. Fitting results for product 3.

Table 3. Values of fitting errors for individual crushing products.

Approximation Function	Approximation Error		
	Product 1	Product 2	Product 3
Weibull's distribution	1.67	7.78	15.25
Log-norm distribution	4.59	3.79	12.92
Logistic distribution	10.26	17.06	2.05

5. Environmental Issues

As previously mentioned, negative impacts of mineral processing (and mining in general) are evident in various aspects. One of the most commonly described seems to be the dust pollution. A wide range of investigations in this area can be found in the literature, particularly those concerning the operation of the mining industry. These include investigations into dust emissions directly from the open pit mines or quarries resulting from their routine operation [36–39], emissions from tailing deposits [40] or transport of run-of-mine and aggregate products, including loading and unloading of material [41]. Operations of mechanical processing are characterized by various rates of dust emission, depending on the size of particle, type of device and its productivity, and crushing stage [42]. Typical relationships between the size of handling material and the total amount of dust generated are described by means of exponential function with negative relationships between the size of the particle and emission volume (Formula (2)):

$$E_{TSP} = \frac{A}{d^B} \tag{2}$$

where E_{TSP}, total emission of dust particles (mg/m^3); d, particle size (mm); A, B, coefficients. Selected values of total dust emissions according to different sources are presented in Table 4.

Table 4. The TSP emission of selected devices according to various investigations.

Processing Stage	Relative Emission (Primary Crushing = 1) [43]	Total Emission [mg/m^3] [44]
Primary crushing	1	2.8
Secondary crushing	3 [45]	3.2
Tertiary crushing	51 (dry), 2 (wet)	30
Screening (dry)	214	No data
Screening (wet)	12	No data

The dust emission is often described as TSP or total suspended particulates. It denotes particles with a diameter smaller than 20 μm, as larger grains usually fall to the ground quickly and do not constitute the air contaminants.

To overcome this negative impact, many models which determine and predict the volume of dust emissions have been developed. These models consider the general principles [46,47] and are built on empirical data relating to the specific site [18,19]. The formulas were devised on the bases of empirical data, the specific location of a mine, and atmospheric conditions. Parameters characterizing properties of the material (such as moisture and silt contents) were taken into account.

Noise emission is another important factor that affects the life of local society around the mineral processing plant, as well the working conditions at the site [50]. The limitation of this source of pollution seems to be relatively less complex technically than in the case of dust, and typically involves building suitable soundproof and sound-absorbing screens and walls.

6. Summary

Mining and mineral processing are relevant for numerous disciplines and have significant impacts on the operation of various sectors of the economy, especially in raw material management within the metal value chain production. An increasing number of mining companies are facing problem concerning decreases in orebody grades and finer mineralization due to the depletion of deposits. At such a time, the idea of zero-waste economy gains importance and popularity. These aspects give rise to the development of mineral processing technology and the more efficient utilization of raw material. Efforts are focusing more on the effective utilization of recent techniques in mechanical processing than on the introduction of brand new feed material treatment technology. This article presented select problems that are especially popular in contemporary mineral processing at the stage of mechanical processing. The increased effectiveness of specific operations can be observed from different scopes, but energy efficiency and useful mineral loss seem to be of major importance. The development of computational techniques and new methods of material analysis will undoubtedly create new opportunities and potential improvements of operation effectiveness.

Not all directions of development have been covered, and the visual analysis and characterization of grained materials are only mentioned. Nevertheless, the results of many investigations and the operational practice of mineral processing plants, show that the stage of the mechanical processing of raw materials preliminarily impacts the potential of effective separation and useful metal recovery.

It is also necessary to highlight major critical aspects and gaps that should be the objects of investigations both in the near future and from a longer-term perspective. There is some consensus that the high energy consumption of comminution processes stands among the most significant issues within the field. However, the problem lies in the more efficient utilization of energy for the breakage and limitation of losses, particularly in fine grinding operations. This is connected to the need for handling the fine mineralized feed material. The scale problem also needs to be solved in electromagnetic grinding, as well as in high-voltage pulse breakage. To some extent, this is also valid for other operations of mechanical processing, such as vibrating mills, SAG and AG grinding, and HPGR. It is also necessary to remember that, when facing the depletion of deposits, (Table 5) it is harder to maintain the overarching aim of mineral processing: achieving a high level of useful mineral recovery.

Table 5. World average grades of selected ore minerals [51,52].

Type of Ore	Unit	Average Grade in 80's	Current Average Grade
Copper	%	1.5	0.62
Lead + Zinc	%	8	6.05
Nickel	%	4	1
Gold (surface mining)	g/Mg	3.5	2

Funding: This research received no external funding.

Conflicts of Interest: The author declares no conflict of interest.

Abbreviations

HPGR	high-pressure grinding rolls
ROM	run-of-mine
CAPEX	capital expenditure
OPEX	operational expenditure
DEM	discrete element method
WOS	Web of Science (database)
PSD	particle size distribution
SEM	scanning electron microscopy
MLA	mineral liberation analyzer
LS	least squares (method)
RRB	Rosin–Rammler–Bennett (distribution)
TSP	total suspended particulates
SAG	semi-autogenous grinding
AG	autogenous grinding

References

1. Gawenda, T. *Zasady Doboru Kruszarek Oraz Układów Technologicznych w Produkcji Kruszyw łamanych. Monography No. 304*; AGH Public House: Cracow, Poland, 2015.
2. Tromans, D. Mineral comminution: Energy efficiency considerations. *Min. Eng.* **2008**, *21*, 613–620. [CrossRef]
3. Fuerstenau, D.W.; Abouzeid, Z.M. The energy efficiency of ball milling in comminution. *Int. J. Min. Proc.* **2002**, *67*, 161–185. [CrossRef]
4. Gawenda, T. The influence of rock raw materials comminution in various crushers and crushing stages on the quality of mineral aggregates. *Miner. Res. Manag.* **2013**, *29*, 53–65. [CrossRef]
5. Gawenda, T. Comparative Analysis of Mobile and Stationary Technological Sets for Screening and Grinding. *Ann. Set Environ. Prot.* **2013**, *15*, 1318–1335.
6. Bearman, K. Step change in context of comminution. *Min. Eng.* **2013**, *43–44*, 2–11. [CrossRef]
7. Morrell, S. Predicting the overall specific energy requirement of crushing, high pressure grinding roll and tumbling mill circuits. *Min. Eng.* **2009**, *22*, 544–549. [CrossRef]
8. Wang, C.; Nadolski, S.; Mejia, O.; Drozdiak, J.; Klein, B. Energy and cost comparison of HPGR based circuits with the SABC circuit installed at the Huckleberry Mine. In Proceedings of the 45th Annual Canadian Mineral Processors Operators Conference, Ottawa, ON, Canada, 22–24 January 2013; pp. 121–135.
9. Klingmann, H.L. HPGR benefits at Golden Queen Soledad Mountain gold heap leaching project. In Proceedings of the Randol Innovative Metallurgy Forum, Perth, Australia, 21–24 August 2005.
10. Yin, W.; Tang, Y.; Ma, T.; Zuo, W.; Yao, J. Comparison of sample properties and leaching characteristics of gold ore from jaw crusher and HPGR. *Min. Eng.* **2017**, *111*, 140–147. [CrossRef]
11. Tang, Y.; Yin, W.Z.; Wang, J.X.; Zuo, W.R.; Cao, S.H. Effect of HPGR comminution scheme on particle properties and heap leaching of gold. *Canadian Metallurg. Quart.* **2020**, *59*, 1–7. [CrossRef]
12. Saramak, D.; Saramak, A. Potential Benefits in Copper Sulphides Liberation through Application of HRC Device in Ore Comminution Circuits. *Minerals* **2020**, *10*, 817. [CrossRef]
13. Baum, W.; Ausburn, K. HPGR comminution for optimization of copper leaching. *Min. Met. Process.* **2011**, *28*, 77–81. [CrossRef]
14. Ghorbani, Y.; Mainza, A.N.; Petersen, J.; Becker, M.; Franzidis, J.-P.; Kalala, J.T. Investigation of particles with high crack density produced by HPGR and its effect on the redistribution of the particle size fraction in heaps. *Min. Eng.* **2013**, *43–44*, 44–51. [CrossRef]
15. Anders, U. Development and prospects of mineral liberation by electrical pulses. *Int. J. Min. Proc.* **2010**, *97*, 31–38. [CrossRef]
16. Gao, P.; Yuan, S.; Han, Y.; Li, S.; Chen, H. Experimental Study on the Effect of Pretreatment with High-Voltage Electrical Pulses on Mineral Liberation and Separation of Magnetite Ore. *Minerals* **2017**, *7*, 153. [CrossRef]
17. Wolosiewicz-Głąb, M.; Ogonowski, S.; Foszcz, D.; Gawenda, T. Assessment of classification with variable air flow for inertial classifier in dry grinding circuit with electromagnetic mill using partition curves. *Physicochem. Probl. Min. Proc.* **2018**, *54*, 440–447.
18. Saramak, D.; Tumidajski, T.; Gawenda, T.; Naziemiec, Z. Ekologiczne aspekty związane z efektami wysokociśnieniowego rozdrabniania w prasach walcowych. *Ann. Set Environ. Prot.* **2013**, *15*, 1580–1593.
19. Phengsaart, T.; Ito, M.; Hamaya, N.; Tabelin, C.B.; Hiroyoshi, N. Improvement of jig efficiency by shape separation, and a novel method to estimate the separation efficiency of metal wires in crushed electronic wastes using bending behavior and "entanglement factor". *Min. Eng.* **2018**, *129*, 54–62. [CrossRef]
20. Ambrós, W.M. Jigging: A Review of Fundamentals and Future Directions. *Minerals* **2020**, *10*, 998. [CrossRef]
21. Stempkowska, A.; Gawenda, T.; Naziemiec, Z.; Ostrowski, K.; Saramak, D.; Surowiak, A. Impact of the geometrical parameters of dolomite coarse aggregate on the thermal and mechanic properties of preplaced aggregate concrete. *Materials* **2020**, *13*, 4358. [CrossRef] [PubMed]
22. Gawenda, T. Układ Urządzeń do Produkcji Kruszyw Foremnych. *Patent PL* **2019**, *233689*, B1. (In Polish)

23. Rosario, P.P. Technical and economic assessment of a non-conventional HPGR circuit. *Min. Eng.* **2017**, *103–104*, 102–111. [CrossRef]
24. Weerasekara, N.S.; Powell, M.S.; Cleary, P.W.; Tavares, L.M.; Evertsson, M.; Morrison, R.D.; Quist, J.; Carvalho, R.M. The contribution of DEM to the science ofcomminution. *Powder Technol.* **2013**, *248*, 3–24. [CrossRef]
25. Tavares, L.M.; Rodrigues, V.; Carles, M.S. Padros, B. An effective sphere-based model for breakage simulation in DEM. *Powder Technol.* **2021**, *392*, 473–488. [CrossRef]
26. Jovanovic, I.; Nikolic, D.; Savic, M.; Zivkovic, Z. Batch composition optimization for the copper smelting process on the example of copper smelter in BOR. *Environ. Eng. Manag. J.* **2016**, *15*, 791–799. [CrossRef]
27. Jovanovic, I.; Savic, M.; Zivkovic, Z.; Boyanov, B.S. An Linear Programming Model for Batch Optimization in the Ecological Zinc Production. *Environ. Model. Assess.* **2016**, *21*, 455–465. [CrossRef]
28. Saramak, D.; Tumidajski, T.; Skorupska, B. Technological and economic strategies for the optimization of Polish electrolytic copper production plants. *Min. Eng.* **2010**, *23*, 757–764. [CrossRef]
29. Naziemiec, Z.; Saramak, D. Application of partition curves in the assessment of mineral products classification processes. *Min. Sci.* **2015**, *22*, 119–129.
30. Lowrison, G.C. *Crushing and Grinding*; Butterworths: London, England, 1974.
31. Brożek, M.; Mączka, W.; Tumidajski, T. *Mathematical Models of Grinding Processes*; AGH Publishing House: Cracow, Poland, 1995. (In Polish)
32. Tumidajski, T.; Saramak, D. *Methods and Models of Mathematical Statistics in Mineral Processing*; AGH Publishing House: Cracow, Poland, 2009; pp. 1–304.
33. Rosin, P.; Rammler, E. The laws governing the fineness of powdered coal. *J. Inst. Fuel* **1933**, *7*, 29–36.
34. Kołomogorow, A.N. Über das logarithmisch normale Verteilungsgesetz der Dimensionen der Teilchen bei Zerstückelung. *Dokl. Akad. Nauk SSSR* **1941**, *31*, 99–101.
35. Nad, A.; Brożek, M. Application of three-parameter distribution to approximate the particle size distribution function of comminution products of dolomitic type of copper ore. *Arch. Min. Sci.* **2017**, *62*, 411–422. [CrossRef]
36. Chakraborty, M.K.; Ahmad, M.; Singh, R.S.; Pal, D.; Bandopadhyay, C. Determination of the emission rate from various opencast mining operations. *Environ. Model. Softw.* **2002**, *17*, 467–480. [CrossRef]
37. Huertas, J.; Camacho, D.; Huertas, M. Standardized emissions inventory methodology for open pit mining areas. *Environ. Sci. Pollut. Res.* **2012**, *19*, 2784–2794. [CrossRef]
38. Chang, C.-T.; Chang, Y.-M.; Lin, W.-Y.; Wu, M.-C. Fugitive dust emission source profiles and assessment of selected control strategies for particulate matter at gravel processing sites in Taiwan. *J. Air Waste Manag. Assoc.* **2012**, *60*, 1262–1268. [CrossRef]
39. Saramak, A.; Naziemiec, Z. Determination of dust emission level for various crushing devices. *Min. Sci.* **2019**, *26*, 45–54. [CrossRef]
40. Cigagna, M.; Dentoni, V.; Grosso, B.; Massacci, G. Emissions of Fugitive Dust from Mine Dumps and Tailing Basins in South-Western Sardinia. In *Mine Planning and Equipment Selection*; Springer: Cham, Switzerland, 2014.
41. Tian, S.; Liang, T.; Li, K. Fine road dust contamination in a mining area presents a likely air pollution hotspot and threat to human health. *Environ. Intern.* **2019**, *128*, 201–209. [CrossRef]
42. Saramak, A. Comparative Analysis of Selected Types of Crushing Forces in Terms of Dust Emission. *Inżynieria Miner. J. Pol. Miner. Eng. Soc.* **2019**, *2*, 151–154.
43. Cecala, A.B.; O'Brien, A.D.; Schall, J.; Colinet, J.F.; Franta, R.J.; Schultz, M.J.; Haas, E.J.; Robinson, J.E.; Patts, J.; Holen, B.M.; et al. *Dust Control Handbook for Industrial Minerals Mining and Processing*, 2nd ed.; National Institute for Occupational Safety and Health: Pittsburgh, PA, USA, 2019.
44. Saramak, D.; Wasilewski, S.; Saramak, A. Influence of copper ore comminution in HPGR on downstream minerallurgical processes. *Arch. Metall. Mater.* **2017**, *62*, 1689–1694. [CrossRef]
45. *Evaluation of Fugitive Dust Emissions from Mining—Report*; PEDCo-Environmental Specialists Inc.: Cincinnati, OH, USA, 1976.
46. Holmes, N.S.; Morawska, L.A. Review of dispersion modeling and its application to the dispersion of particles: An overview of different dispersion models available. *Atmos. Environ.* **2006**, *40*, 5902–5928. [CrossRef]
47. Prakash, J.; Singh, G.; Pal, A.K. Air pollution dispersion modeling performance for mining complex. *Environ. We Int. J. Sci. Technol.* **2010**, *5*, 205–222.
48. Chaulya, S.K.; Chakraborty, M.K.; Ahmad, M.; Singh, R.S.; Bondyopadhay, C.; Mondal, G.C.; Pal, D. Development of empirical formulae to determine emission rate from various opencast coal mining operations. *Water Air Soil Pollut.* **2002**, *140*, 21–55. [CrossRef]
49. Chaulya, S.K.; Ahmad, M.; Singh, R.S.; Bandopadhyay, L.K.; Bondyopadhay, C.; Mondal, G.C. Validation of two air quality models for Indian mining conditions. *Environ. Monit. Assess.* **2002**, *82*, 23–43. [CrossRef] [PubMed]
50. Saramak, A.; Naziemiec, Z.; Saramak, D. Analysis of noise emission for selected crushing devices. *Min. Sci.* **2016**, *23*, 145–154. [CrossRef]
51. Saramak, D. Simulation of comminution effects in HPGR for the feed material variable content of fine particles. *Miner. Res. Manag.* **2015**, *31*, 123–136.
52. Calvo, G.; Mudd, G.; Valero, A.; Valero, A. Decreasing Ore Grades in Global Metallic Mining: A Theoretical Issue or a Global Reality? *Resources* **2016**, *5*, 36. [CrossRef]

Article

Application of Wavelet Filtering to Vibrational Signals from the Mining Screen for Spring Condition Monitoring

Natalia Duda-Mróz, Sergii Anufriiev * and Paweł Stefaniak

KGHM Cuprum Research and Development Centre Ltd., Gen. W. Sikorskiego 2-8, 53-659 Wroclaw, Poland; natalia.duda@kghmcuprum.com (N.D.-M.); pawel.stefaniak@kghmcuprum.com (P.S.)
* Correspondence: sergii.anufriiev@kghmcuprum.com

Abstract: The main task of mineral processing plants is to further process the raw material extracted in the mining faces into a concentrate with the highest possible concentration of the final product. In practice, it is a complex process in which several stages can be distinguished. After the ore has been transported to the surface by the skip shaft, one of the first steps is sieving the ore, which is typically performed using vibrating mining screens. In a typical Ore Enrichment Plant, the screening process is carried out by several such machines. This is a typical bottleneck in the technological chain. For this reason, the main challenge for users is to achieve the highest reliability and efficiency of these technical facilities. The solution is to focus on predictive maintenance strategies based on the development of monitoring and advanced diagnostic procedures capable of estimating the time of safe operation. This work was developed as part of an advanced diagnostic system ensuring comprehensive technical conditioning and early fault detection of components such as the engine, transmission, bearings, springs, and screen. This article focuses on vibration data. The problem of damage detection in the presence of periodically impulsive components resulting from falling feed material on the screen and its further screening process has been considered. These disturbances are of a non-Gaussian noise nature, the elimination of which is essential to extract the fault-related signal of interest. One solution may be to properly smooth and filter the raw signal. In this article, a wavelet filtering technique is applied. First, the wavelet filtering procedure is described. In the next step, the performance of a wavelet filter is investigated depending on its parameters. Then, the results of wavelet filtering are compared with such methods as low-pass filtering and smoothing using a moving average. Finally, the impact of wavelet filtering on the calculation of screen trajectories is investigated.

Keywords: mineral processing; sieving screen; diagnostics; predictive maintenance; wavelet transformation

Citation: Duda-Mróz, N.; Anufriiev, S.; Stefaniak, P. Application of Wavelet Filtering to Vibrational Signals from the Mining Screen for Spring Condition Monitoring. *Minerals* **2021**, *11*, 1076. https://doi.org/10.3390/min11101076

Academic Editors: Daniel Saramak, Marek Pawełczyk and Tomasz Niedoba

Received: 22 July 2021
Accepted: 26 September 2021
Published: 30 September 2021

1. Introduction

Currently, in the mining sector, there is a rapid development of technology ensuring operational supervision of processes and technical facilities based on online monitoring. A typical mining enterprise wishing to remain competitive in the mineral resources market must plan the production process, maintenance, and materials management in advance based on real data. If we look at machinery systems as a network of interconnected vessels, the bottlenecks are one of the most critical. In order to avoid downtime and to carry out maintenance works in a controlled manner, it is crucial to develop advanced diagnostic systems [1]. The key is to provide appropriate sensors that significantly exceed the human senses but also to develop advanced procedures for the extraction of fault-oriented features, procedures for inferencing the diagnostic state of individual components, and procedures for estimating the residual lifetime. Currently, predictive maintenance is developed strongly based on the assumptions of the Industrial Internet of Things technologies, in which the sensed objects are connected to the Internet and can communicate with each other and the superior system [2–7].

In this article, one of the critical technical objects of the Ore Enrichment Plant—the vibrating screen used for the sieving process—is considered. The main purpose of sieving is to separate the fine fraction (suitable for milling) from the coarse fraction (requiring crushing). The horizontal schema of the vibrating screen is shown in Figure 1.

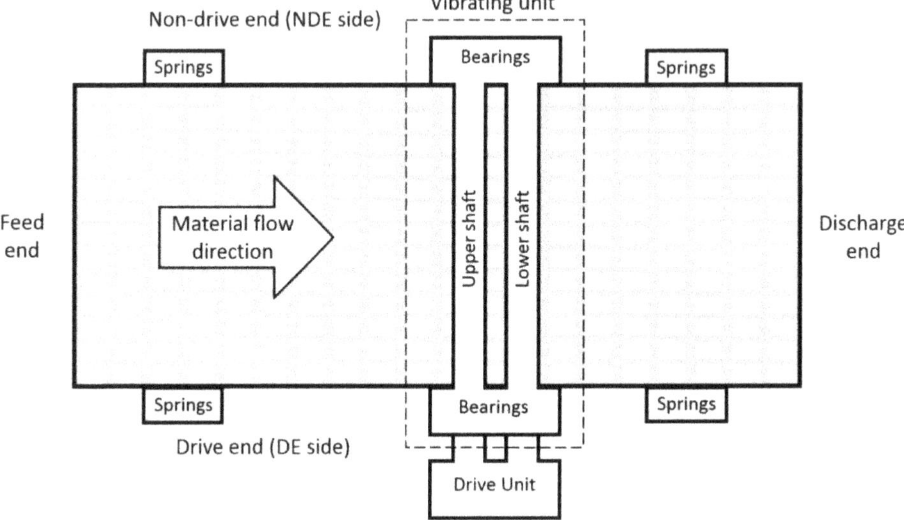

Figure 1. The schema of the vibrating screen.

This process is the first stage to which the ore is subjected after direct transport to the surface, precisely before crushing and milling. At this stage, most of the metal components (e.g., anchors) are also captured using an electromagnet. The research was conducted in one of the Polish underground copper ore mines of KGHM, where the vibrating screens are located close to a mine shaft. Depending on the mining plant, the mesh size in the screen ranges from 20 to 40 mm. In all mines, the number of these objects is very limited (Polkowice—3, Lubin—3, Rudna—6); therefore, the vibrating screen is one of the typical bottlenecks in the entire processing plant. In practice, there are various types of vibrating screens. We can distinguish, among others, rotating vertical cylindrical screens and vibrating horizontal linear screens. The ore separation process in terms of grain size is based on induced vibrations. These vibrations are generated by one or two rotating shafts of unbalanced masses driven by drive units with belt transmission. A standard multi-deck vibrating screen is made of the main box and side plates which are connected by transverse reinforcing beams, upper and lower sieving decks, a deck under the screen, and multiple springs as supports. The feed is transferred to the plates by the conveyor and a dispenser located above the screen (Figure 2) [8].

The current demands of users of these devices are related to the development of monitoring systems and early response tools to the development of damage to components such as the engine, transmission, bearings, springs, or the screen itself [9]. Analyzing the literature, you can find many monitoring systems available on the market designed for processing machines; some of them are strictly dedicated to vibrating screens. Most of the commercial solutions [10–14] concern rolling bearings diagnostics based on vibration signals. Leading manufacturers also offer the expansion of the system with additional temperature sensors and lubrication oil particle (i.e., wear debris) counters. This device is exceptionally specific from a diagnostic point of view due to non-stationary operating conditions but also due to the presence of random high-impulse disturbances resulting from the grains bumping into the screen structure. If we do not separate the non-informative impulses from the informative pulses, the further diagnostic process will not be reliable. This issue has been strongly developed as a part of the project financed from EU funds under the acronym OPMO (Operation monitoring of mineral crushing machinery, [15]), which aims to build an advanced diagnostic system for selected mineral processing assets [16]. In this article, we will focus on the problem of diagnosing supporting springs that play a key role in the sieving process. Critical is their stiffness, which determines the effective operation of the entire screen. In the case of steel springs, a linear deformation characteristic up to a certain degree of deformation is observed. As the time of operation progresses, the stiffness of the springs is lost, which affects the amplitude and frequency of vibrations. Over time, they break as a result of high-cycle fatigue (HCF). Due to their complicated geometry and continuous movement during the screen operation, the number of possible diagnostic methods is limited. The optimal source of data can be vibration data, not only for the diagnosis of flexible springs but also for rolling bearings. One approach may be to plot orbits from two orthogonal vibration signals. Given this representation of the spring motion, it is possible to identify anomalous spring behavior with the knowledge of the correct operating parameters. In [8], the authors presented the main assumptions of such an approach. They described the dynamic model of the vibrating screen and paid special attention to the stochastic influence of the feed on the disturbance of the vibration signal, especially impacts from material falling on the upper deck and impacts from the material inside the screen. They also proposed a spring failure simulation procedure. The method was further developed in [17]. In the case of diagnosing the vibrating screen state, the key is to take into account two phenomena: time-varying load (due to the variable amount of feed on decks) as well as impulsive load considered to be impulsive background noise. In this article, we focus on the problem of signal denoising from these non-informative components. So far, this problem has been solved mainly in the field of diagnostics of rolling bearings, where the important issue is the detection of the cyclic impulse signal in the presence of non-cyclic impulse noise [18–21].

Figure 2. The investigated technical object—vibrating screen: (**a**) view of the whole device; (**b**) feed; (**c**) upper sieving deck.

In this article, a wavelet filtering technique [22–37] was applied to vibrational signals collected in the ore processing plant from a mining screen. The main purpose was to compare the results of wavelet filtering to simple smoothing and filtering techniques, paying special attention to the difference in trajectory calculations. The structure of the article is as follows: In Section 2, the input data and the methods used for signal processing are described. In Section 3, all the main results are presented. First, the wavelet filtering is compared to moving-average and low-pass filters on raw signals. Then, the trajectory of the screen is calculated for the raw signal, low-pass-filtered signal, and wavelet-filtered signal. In Section 4, the obtained results are discussed, and our conclusions are drawn.

2. Materials and Methods

2.1. Input Data Description

An overall description of the diagnostic system installed on the investigated mining screen can be found in [18]. The vibrational data from the accelerometers were recorded with a sampling frequency of 16 or 48 kHz. There were 16 accelerometers, one vertical and one horizontal, for each of four bearings and four spring sets. In order to reduce the amount of stored data, a 1 min recording was taken every 15 min. An example of raw data is shown in Figure 3. Based on these data, the trajectory of the screen, which is a key characteristic for screen diagnostics and maintenance, as well as other diagnostic features, can be calculated.

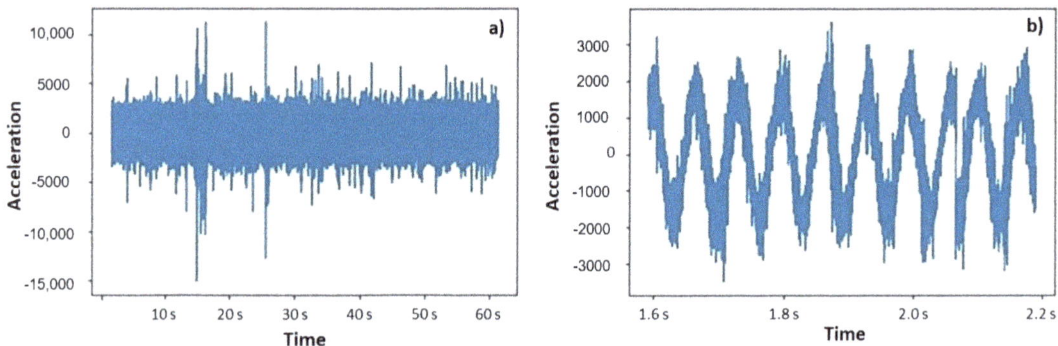

Figure 3. Vibrational signal recorded from a single accelerometer installed on the vibrating screen during (**a**) one-minute-long measurement session (**b**) and a close-up look.

On the left plot, one can see the whole signal recorded during a single measurement. Some significant excitation is observed around the fifteenth second of the measurement, which probably corresponds to a large piece of ore falling onto the screen. Such rocks can be detected using computer vision and audio processing technics based on recorded video signals of the incoming ore flow [22]. Based on the results of this detection, such excitation can be filtered out, which is planned in the future but has not been performed in this article due to the unavailability of video data for the studied period. On the right, a closer look at the same signal is shown, on which natural vibrations of the screen are visible.

The first step of data processing was the transformation of electric signals from the accelerometers into SI units (m/s^2). The following formulas were used for springs:

$$a = \frac{0.1733 \cdot L}{S \cdot g} \tag{1}$$

where L is the measured value from the accelerometer, S is the sensitivity of the accelerometer ($S = 100$mV/g), and $g = 9.81$ m/s^2 is the standard gravity.

The Fourier transformation of the signal for a spring set is shown in Figure 4. In Plot (a), the whole spectrum is shown. There is one dominating frequency of about 15 Hz, which corresponds to the rotation frequency of the shafts. Besides, some excitations are observed in the frequency range of 150–10,000 Hz, which are visible clearly in Plot (c). Generally, the amplitude of these excitations is much lower than the amplitude of the screen's working frequency. For frequencies higher than 10 kHz, almost no excitations are observed, although this range could be important for bearings diagnostics.

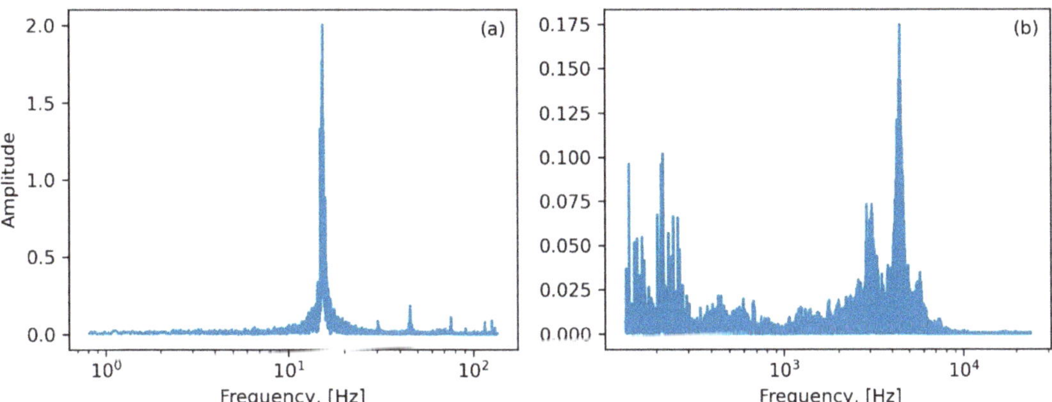

Figure 4. The frequency spectrum of the acceleration signal: (**a**) low-frequency range; (**b**) high-frequency range.

2.2. Wavelet Denoising Procedure

In this subsection, the procedure of wavelet filtering is described. The wavelet denoising procedure consists of three main steps (for more detail, read [23–38]):

- Multilevel wavelet decomposition.
- Finding thresholds for detail coefficients.
- Reconstruction of the signal.

Wavelet transform is a tool that cuts up the signal into detail coefficients (C_D), approximation coefficients (C_A), and downsamples (Figure 5). The detail coefficients can be defined as high-frequency coefficients $y_{high}[n] = \sum\limits_{i=-\infty}^{\infty} s[i]h[2n-i]$, and the approximation coefficients can be defined as low-frequency coefficients $y_{low}[n] = \sum\limits_{i=-\infty}^{\infty} s[i]g[2n-i]$, where i is a sampling data point, n is the size of the sampling data, $s[i]$ is the raw signal, and $g[2n-i]$ and $h[2n-i]$ are low-pass and high-pass filters. The wavelet function is composed of the scaled and translated copies of the scaling function $\phi(x) = \sum\limits_{n} h(n)\sqrt{2}\phi(2x-n)$ and the mother wavelet function $\psi(x) = \sum\limits_{n} g(n)\sqrt{2}\phi(2x-n)$.

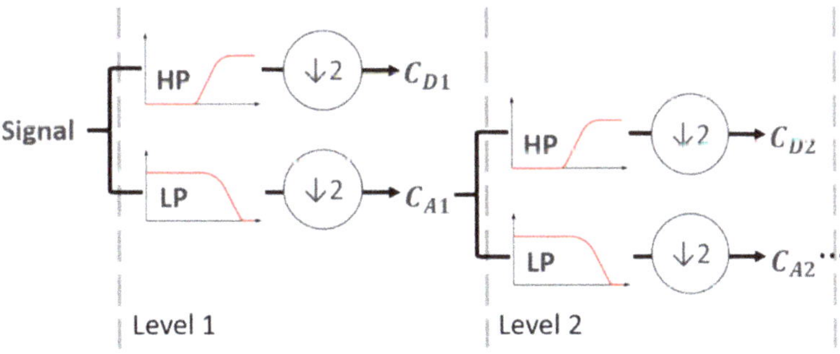

Figure 5. A conceptual schema of a multilevel wavelet decomposition procedure.

After signal decomposition into detail and approximation coefficients, it is necessary to threshold detail coefficients. One of the methods is hard thresholding, described by equation $h_\lambda(x) = x \times 1_{\{|x|>\lambda\}}$. The signal x remains unchanged if its values are lower than $-\lambda$ or greater than threshold λ; otherwise, the values are replaced with zeros. Another method is soft thresholding $s_\lambda(x) = \text{sign}(x) \times \max(|x| - \lambda, 0)$. Here, values greater in magnitude than the threshold are shrunk towards zero by subtracting the threshold from it. The method that is the combination of both methods explained is semi-soft thresholding, described by the equation below:

$$c_\lambda(x) = \begin{cases} x, & |x| > \lambda_2, \\ \text{sign}(x) \times \frac{\lambda_2(|x|-\lambda_1)}{\lambda_2-\lambda_1}, & \lambda_1 \leq |x| \leq \lambda_2, \\ 0, & |x| < \lambda_1. \end{cases}$$

Thresholds are calculated for each detail coefficient d_j (i.e., noise). A common threshold choice is, for example, universal threshold $\lambda = \sigma\sqrt{2\log n}$, where n is the length of noise and σ is the median of absolute values from noise divided by 0.6745 [39]. Our threshold choice was $\lambda_1 = \mu + 2\sigma^2$ and $\lambda_2 = \mu + 3\sigma^2$, where μ is the mean and σ^2 is the variance of the detail coefficient (Figure 6).

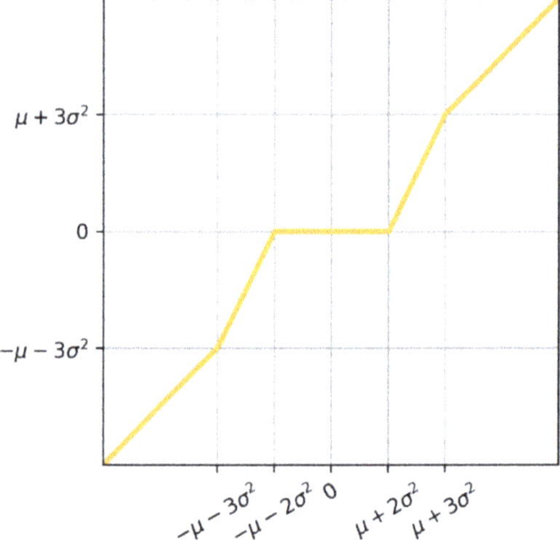

Figure 6. Semi-soft thresholding.

Other more widely used thresholding methods are adaptive threshold selection, using the principle of Stein's Unbiased Risk Estimate (SURE) [40], and the minimax threshold [40]. (For more about thresholding, read [40–42].)

After thresholding, the signal can be reconstructed by inverse multilevel wavelet transform. The noise is the difference between the raw signal and the denoised signal.

2.3. Trajectory Calculation Procedure

One of the most important diagnostic characteristics of the mining screen is its trajectory (also referred to as orbit). The trajectory of the mining screen can be determined by double integration of the original acceleration signal in vertical and horizontal directions. The processing steps for the vertical vibrations and the horizontal vibrations are shown in Figure 7. At the start, the input is the acceleration of the horizontal or vertical vibrations. The algorithm must be used separately for both directions. Next, outputs must be merged by time vectors to find trajectory.

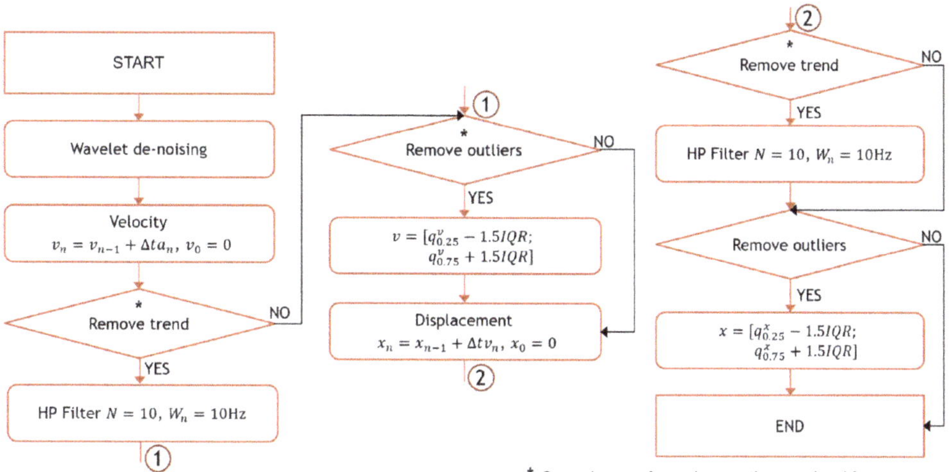

Figure 7. Procedure for determining the displacement based on a one-dimensional vibration acceleration signal (horizontal or vertical).

The first step was denoising the signal. On the schema, the wavelet denoising is suggested, but some other methods and calculations without filtering at all will be compared later in this article. After wavelet denoising, the signal was integrated with the use of the Euler method, and the velocity was obtained. Then, the velocity values were filtered by a 10-order high-pass Butterworth filter to remove slowly changing trends in data, such as gravity. Significant outliers may appear due to possible breaks in the recorded acceleration signal. This is why Tukey's fences technique [37] was applied to remove outliers. The signal was integrated once more with the use of the Euler method, and the coordinates of the screen were obtained. The trend and outliers in the coordinates were removed in the same way as in velocity signals.

After all the horizontal and vertical coordinates are connected by time vectors, the 2D vector of the position changed in time can be obtained for any period of time. The expected trajectory of the screen is an ellipse. For long-term trajectory analysis, the calculation of some key parameters of the ellipse can be useful. In order to achieve this, an ellipse equation was fitted to the trajectories obtained numerically. The following total least squares estimator for 2D ellipses was used:

$$\begin{cases} x_t = x_c + a \cos\theta \cos t - b \sin\theta \sin t \\ y_t = y_c + a \sin\theta \cos t + b \cos\theta \sin t \end{cases} \tag{2}$$

$$d = \sqrt{(x - x_t)^2 + (y - y_t)^2}, \tag{3}$$

where (x_t, y_t) is the closest point on the ellipse to (x, y). Thus, d is the shortest distance from the point to the ellipse. By minimizing d, the parameters of the elliptical trajectory can be calculated.

3. Results

In this section, the main results are shown. The results are presented for a selected spring set of a sieving screen.

The mother wavelet and level of decomposition were selected based on the Pearson correlation coefficient between the raw and denoised signals. The different methods of thresholding have been checked. Satisfactory results have been obtained for the biorthogonal 2.2 wavelet (bior2.2) with the sixth level of decomposition and the semi-soft thresholding method.

First, an example of filtering is shown, then the wavelet filtering is compared to some other method, and finally, the orbit calculation results are compared. In this article, we focus on the signals for springs, although similar analyses could be carried out for bearings. (See [26] for an interesting approach for bearings.) The wavelet filter was applied to the vibrational signal from the spring set of the sieving screen. In Figure 8, the signals before and after the wavelet denoising, as well as the noise, are presented.

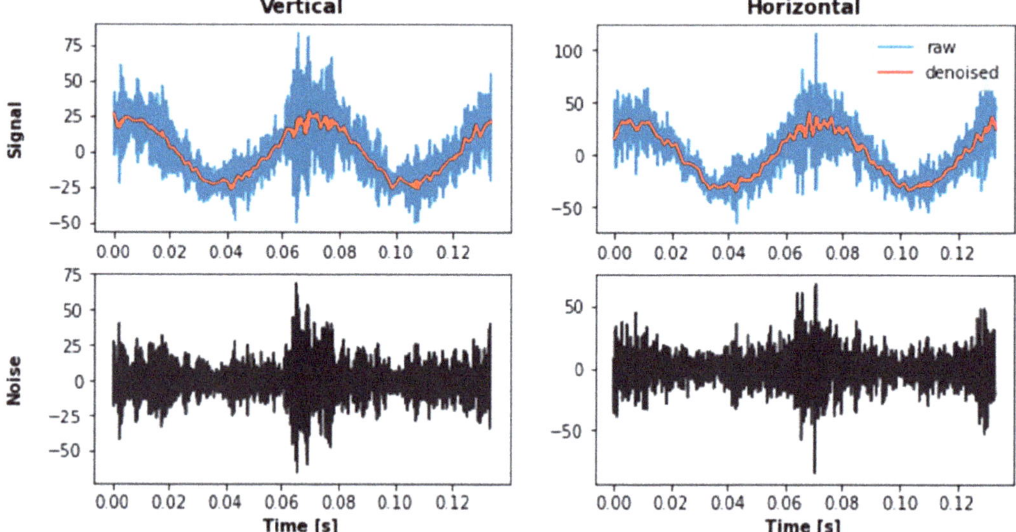

Figure 8. Wavelet denoising of the vibrational signal recorded on the spring.

In the first step of wavelet denoising, the acceleration signal was put into cascaded filters (see Figure 5) with the biorthogonal 2.2 wave. The bior2.2 wave is symmetric, not orthogonal, and biorthogonal. Next, for each detail coefficient, the semi-soft thresholding method was applied (Figure 6).

A key parameter that affects the performance of the wavelet filtering is the level of the filter, i.e., how many steps are performed in the procedure, shown in Figure 5. In Figure 8, a denoise level equal to 6 was chosen. In Figure 9, the denoising results obtained with different levels of wavelet decomposition are compared. As seen for lower values (n = 3), a significant amount of noise is present in the filtered signal. On the other hand, for higher values, either some details of the original signal are lost (n = 9), or the signal is distorted completely (n = 12). We have chosen an intermediate value (n = 6), which gives a smooth enough signal but does not distort it at the same time.

Next, the results of the wavelet filtering are compared with some other techniques of signal processing. In Figure 10, the wavelet denoising is compared with a moving-average filter (AVG), which is one of the obvious choices for smoothing data. The window size of the moving average was equal to 150 observations, which correspond to 0.003125 s. In general, wavelet filtering improves the signal-to-noise ratio of the original signal. It usually fits within the range of the original signal and, at the same time, provides a rather smooth line without small excitations. Slower changes in the signal are caught, while the faster excitations are filtered out. Although the output of the moving-average filter is also very smooth, it does not fit the original signal as well as the wavelet signal. In the case of larger excitations observed in the original signal, some major deviations between the original signal and the moving-average signal are observed. Those deviations seem to be small, but during further processing steps (for example, orbit calculations), such small errors can accumulate and lead to distorted results and inappropriate screen diagnostics.

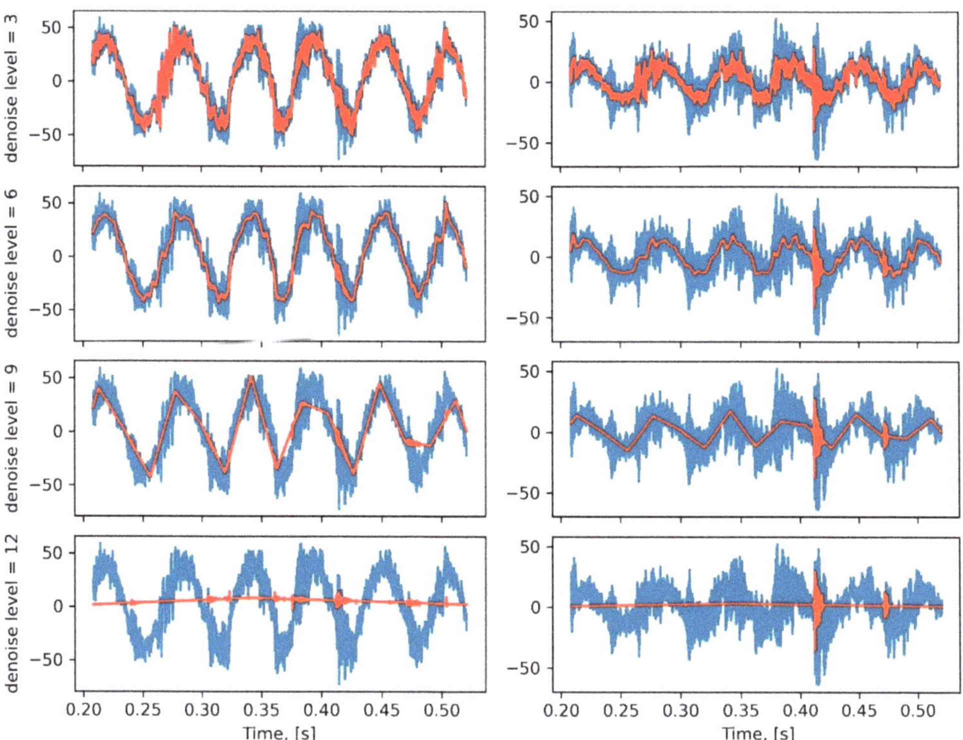

Figure 9. Comparison of different levels of signal denoising using a multilevel wavelet transformation. Blue line shows the original signal; red line shows the filtered signal.

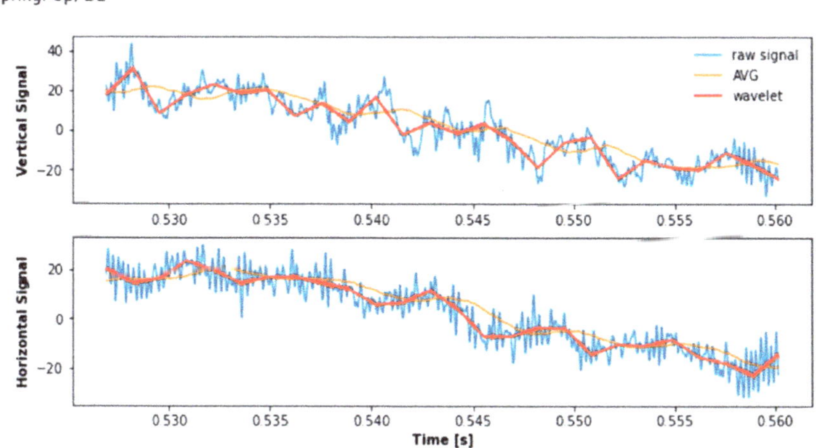

Figure 10. Comparison of wavelet filtering to signal smoothing using the moving average filter for vertical and horizontal orientations of accelerometers.

Next, we compared the results of wavelet filtering with a low-pass filter. The original signal contains very slow excitation, corresponding to the movement of the whole screen with a frequency of 15 Hz. An obvious way to isolate these vibrations is to apply a low-pass filter. The results of the comparison are shown in Figure 11. The wavelet denoising is compared to a two-order low-pass Butterworth filter with a cutoff frequency equal to 18 Hz. The filter must be applied twice: once forward and once backward. The combined filter has a zero phase and a filter order twice that of the original. First of all, the results obtained by applying the low-pass filter are very smooth because, basically, it is a single harmonic excitation. On the other hand, the amplitude of the signal after low-pass filtering is lower, which can result in appropriate screen diagnostics. The wavelet results can be improved by smoothing or adaptive changing of the level or types of the wave.

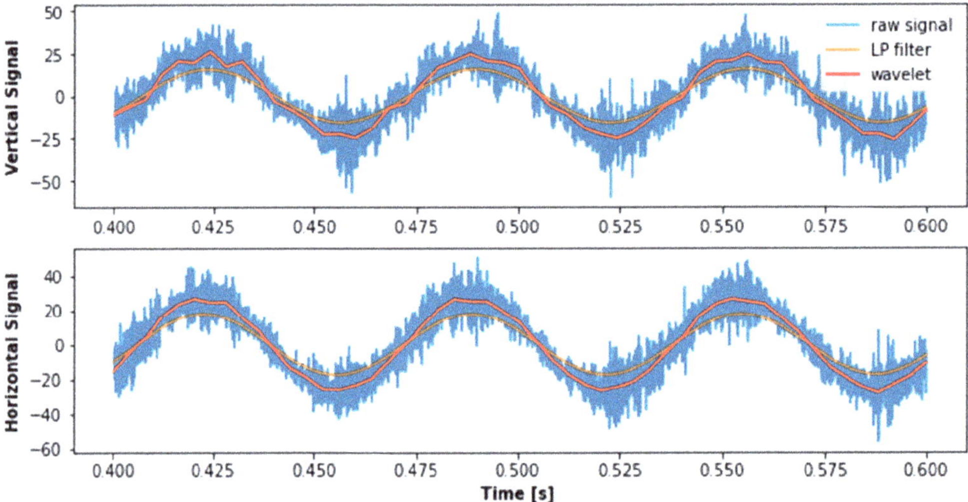

Figure 11. Comparison of wavelet filtering to low-pass filtering for vertical and horizontal orientations of accelerometers.

Finally, the results obtained by filtering the original signal using different techniques were used to calculate the trajectory of the sieving screen. The trajectory was calculated using the procedure described before in Figure 6 using the raw signal, the signal processed by low-pass filtering, and the signal processed by wavelet filtering. The results are shown in Figure 12. The trajectory calculated on the raw signal is shown in blue, the one calculated after low-pass filtering in orange, and the one filtered by wavelets in red. Although the results are similar, some major differences may be observed. The main issue with the trajectory calculated based on the raw signal is the fact that it is diverging fast. The difference between the two rotations is significant, and it increases further with time. It is even more visible on a magnified plot. This can be explained by the accumulation of errors during the integration of the original signal. Both filtering techniques deal with this issue well: the divergence is much lower for both the low-pass and wavelet filters. Lower divergence of the trajectory will allow the use of longer parts of the signal for trajectory calculation, which will allow for the calculation of the parameters of a trajectory in a more reliable way.

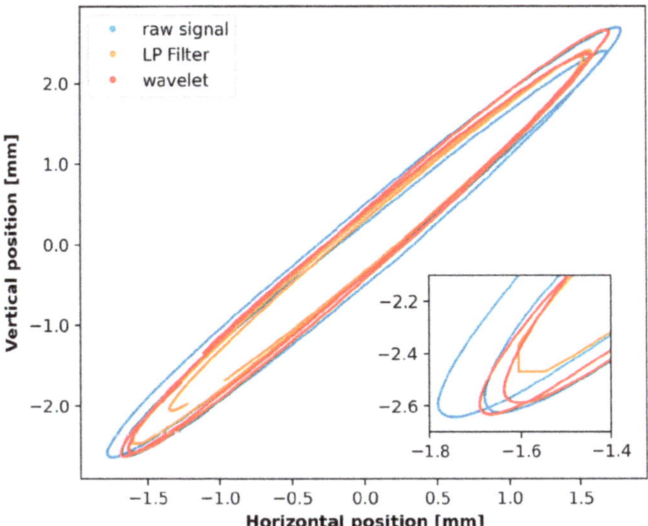

Figure 12. Orbits calculated using different kinds of data filtering techniques: no filtering (blue line), low-pass filtering (orange line), and wavelet filtering (red line).

When it comes to a comparison between the results of the low-pass filter and the wavelet filter, the main difference is the magnitude of the trajectory. The length of the trajectory in the case of low-pass filtering is significantly lower than in the case of wavelet filtering. This fact can lead to inappropriate diagnostics of the screen. Many methods for predictive maintenance rely on setting the alarm threshold for such parameters as the length of the trajectory. In the case of calculations using the low-pass filter, the measured parameters of the orbit may be lowered, which can potentially lead to unexpected failures due to the fact that the maximum safe magnitude was exceeded.

4. Discussion and Conclusions

In this paper, a novel procedure for technical condition monitoring of a vibrating screen in the presence of impulsive noise has been presented. The methods found in the literature are mainly dedicated to noise reduction for the purpose of detecting local damage to rolling bearings. In this paper, we have focused on springs. One of the popular methods of assessing the technical condition of springs is to analyze the trajectory of their movement. In order to estimate orbits, it is necessary to have access to two orthogonal vibration signals. Unfortunately, the presence of impulsive background noise due to large pieces of ore falling down distorts these estimates. One solution may be to properly smooth and filter the raw signal.

In this article, a wavelet filtering technique is applied for this purpose. The filtering procedure, as well as the procedures for orbit calculation and parametrization, were described. Then, these procedures were applied to vibrational data collected in an ore processing plant from a machine operating in industrial conditions. The results of wavelet filtering are compared with other methods, such as moving-average filtering and low-pass filtering. Next, the trajectory was calculated using different preprocessing techniques, and the results were compared.

Wavelet filtering has shown some improvement compared to both moving-average and low-pass filtering. When compared with the moving-average filter, the wavelet filtering better represents the original signal. The moving average was very sensitive to outliers in the original signal, while the wavelet-filtered signal was, in most cases, within the original signal. Compared to the low-pass filter, the wavelet filter better conserves the magnitude of the original signal.

When it comes to trajectory calculation, preprocessing using wavelet filtering also gave some advantages. Compared to the calculation based on unfiltered data, better convergence of the trajectory was obtained. It will allow for the use of more data for parametrization of the trajectory (fitting and calculation of an ellipse equation to numerically calculated trajectory), which will give more reliable results. On the other hand, compared to low-pass filtering, the magnitude of the orbit calculated using wavelet filtering is closer to the original one, which may allow one to avoid exceeding the working parameters of the screen.

Of course, wavelet filtering has some downsides. First of all, an appropriate decomposition scaling function and the level of the filter need to be chosen properly, which can be time-consuming. However, after optimization of these parameters, wavelet filtering gives good results compared to some other smoothing and filtering techniques. In the future, it is planned to compare the results with some more sophisticated filtering techniques.

Author Contributions: Conceptualization, N.D.-M., S.A. and P.S.; Data curation, N.D.-M. and S.A.; Formal analysis, P.S.; Investigation, S.A. and P.S.; Methodology, N.D.-M., S.A. and P.S.; Project administration, S.A. and P.S.; Software, N.D.-M. and S.A.; Supervision, S.A. and P.S.; Validation, S.A. and P.S.; Visualization, N.D.-M., S.A. and P.S.; Writing—original draft, N.D.-M., S.A. and P.S.; Writing—review and editing, S.A. and P.S. All authors will be informed about each step of manuscript processing, including submission, revision, revision reminder, etc., via emails from our system or the assigned Assistant Editor. All authors have read and agreed to the published version of the manuscript.

Funding: This research was funded by EIT RawMaterials GmbH under Framework Partnership Agreement No. 18253 (OPMO—Operation monitoring of mineral crushing machinery).

Data Availability Statement: Restrictions apply to the availability of these data. Data was obtained from KGHM Polska Miedź S.A. and are available from the authors with the permission of KGHM Polska Miedź S.A.

Conflicts of Interest: The authors declare no conflict of interest. The funders had no role in the design of the study; in the collection, analyses, or interpretation of data; in the writing of the manuscript; or in the decision to publish the results.

References

1. Lanke, A.A.; Hoseinie, S.H.; Ghodrati, B. Mine production index (MPI)-extension of OEE for bottleneck detection in mining. *Int. J. Min. Sci. Technol.* **2016**, *26*, 753–760. [CrossRef]
2. Gackowiec, P.; Podobińska-Staniec, M. IoT platforms for the Mining Industry: An Overview. *Inżynieria Miner.* **2019**, *21*, 530–550.
3. Kruczek, P.; Gomolla, N.; Hebda-Sobkowicz, J.; Michalak, A.; Śliwiński, P.; Wodecki, J.; Stefaniak, P.; Wyłomańska, A.; Zimroz, R. Predictive maintenance of mining machines using advanced data analysis system based on the cloud technology. In *Proceedings of the 27th International Symposium on Mine Planning and Equipment Selection-MPES, Santiago, Chile, 20-22 November, 2018*; Springer: Cham, Switzerland, 2019; pp. 459–470.
4. Gubbi, J.; Buyya, R.; Marusic, S.; Palaniswami, M. Internet of Things (IoT): A vision, architectural elements, and future directions. *Future Gener. Comput. Syst.* **2013**, *29*, 1645–1660. [CrossRef]
5. Dong, L.; Mingyue, R.; Guoying, M. Application of internet of things technology on predictive maintenance system of coal equipment. *Procedia Eng.* **2017**, *174*, 885–889. [CrossRef]
6. Molaei, F.; Rahimi, E.; Siavoshi, H.; Afrouz, S.G.; Tenorio, V. A comprehensive review on internet of things (IoT) and its implications in the mining industry. *Am. J. Eng. Appl. Sci.* **2020**, *13*, 499–515. [CrossRef]
7. Schlemmer, G. *Principles of Screening and Sizing*; Quarry Acadamy: San Antonio, TX, USA, 2016.
8. Krot, P.; Zimroz, R. Methods of Springs Failures Diagnostics in Ore Processing Vibrating Screens. In *IOP Conference Series: Earth and Environmental Science*; IOP Publishing: Bristol, UK, 2019; Volume 362, p. 012147.
9. Kahraman, M.M.; Rogers, W.P.; Dessureault, S. Bottleneck identification and ranking model for mine operations. *Prod. Plan. Control.* **2020**, *31*, 1178–1194. [CrossRef]
10. SKF—Evolution Technology Magazine from SKF Fault Detection for Mining and Mineral Processing Equipment. 2001. Available online: http://evolution.skf.com/fault-detection-for-mining-and-mineral-processing-equipment-2/ (accessed on 22 July 2021).
11. Elmodis—Tool Created for Industry Official Website of the Manufacturer of Diagnostic Equipment. 2020. Available online: https://elmodis.com/en/# (accessed on 1 July 2021).
12. Metso—ScreenWatch®Screen Condition Monitoring Brochure. 2020. Available online: https://pdf.directindustry.com/pdf/metso-corporation/screenwatch-screen-condition-monitoring-brochure/9344-774332.html (accessed on 22 July 2021).

13. Schaeffler—FAG SmartCheck Machinery Moniroting for Every Machine. 2012. Available online: https://www.schaeffler.com/remotemedien/media/_shared_media/08_media_library/01_publications/schaeffler_2/tpi/downloads_8/tpi_214_en_us.pdf (accessed on 22 July 2021).
14. SchenckPeocess—CONiQ®Condition Monitoring in Mineral Processing. The Power to Predict! 2020. Available online: https://www.schenckprocess.com/data/en/files/513/bvp2135en.pdf (accessed on 22 July 2021).
15. OPMO: Operation Monitoring of Mineral Crushing Machinery EIT Raw Materials Website. 2020. Available online: https://eitrawmaterials.eu/project/opmo/ (accessed on 22 July 2021).
16. Krot, P.; Zimroz, R.; Michalak, A.; Wodecki, J.; Ogonowski, S.; Drozda, M.; Jach, M. Development and Verification of the Diagnostic Model of the Sieving Screen. *Shock. Vib.* **2020**, *2020*, 1–14. [CrossRef]
17. Gąsior, K.; Urbańska, H.; Grzesiek, A.; Zimroz, R.; Wyłomańska, A. Identification, decomposition and segmentation of impulsive vibration signals with deterministic components—A sieving screen case study. *Sensors* **2020**, *20*, 5648. [CrossRef] [PubMed]
18. Cai, Z.; Xu, Y.; Duan, Z. An alternative demodulation method using envelope-derivative operator for bearing fault diagnosis of the vibrating screen. *J. Vib. Control.* **2018**, *24*, 3249–3261. [CrossRef]
19. Nowicki, J.; Hebda-Sobkowicz, J.; Zimroz, R.; Wylomanska, A. Local Defect Detection in Bearings in the Presence of Heavy-Tailed Noise and Spectral Overlapping of Informative and Non-Informative Impulses. *Sensors* **2020**, *20*, 6444. [CrossRef]
20. Wodecki, J.; Michalak, A.; Zimroz, R. Local damage detection based on vibration data analysis in the presence of Gaussian and heavy-tailed impulsive noise. *Measurement* **2021**, *169*, 108400. [CrossRef]
21. Skoczylas, A.; Anufriiev, S.; Stefaniak, P. Oversized ore pieces detection method based on computer vision and sound processing for validation of vibrational signals in diagnostics of mining screen. *SGEM* **2020**, *20*, 829–839. [CrossRef]
22. Aminghafari, M.; Cheze, N.; Poggi, J. Multivariate denoising using wavelets and principal component analysis. *Comput. Stat. Data Anal.* **2006**, *50*, 2381–2398. [CrossRef]
23. Dautov, C.P.; Ozerdem, M.S. Wavelet transform and signal denoising using Wavelet method. In Proceedings of the 26th Signal Processing and Communications Applications Conference (SIU), Izmir, Turkey, 2–5 May 2018.
24. Jiang, J.; Guo, J.; Fan, W.; Chen, O. An Improved Adaptive Wavelet Denoising Method Based on Neighboring Coefficients. In Proceedings of the 8th World Congress on Intelligent Control and Automation, Jinan, China, 6–9 July 2010.
25. Jha, R.K.; Swami, P.D.; Singh, D. Comparison and Selection of Wavelets for Vibration Signals Denoising and Fault Detection of Rotating Machines Using Neighborhood Correlation of SWT Coefficients. In Proceedings of the International Conference on Advanced Computation and Telecommunication (ICACAT), Bhopal, India, 28–29 December 2018.
26. Qin, Z.; Chen, L.; Bao, X. Wavelet Denoising Method for Improving Detection Performance of Distributed Vibration Sensor. *IEEE Photonics Technol. Lett.* **2012**, *24*, 542–544. [CrossRef]
27. Gao, H.Y.; Bruce, A.G. Waveshrink with Firm Shrinkage. *Stat. Sin.* **1997**, *7*, 855–874.
28. Galiana-Merino, J.J.; Rosa-Herranz, J.; Giner, J.; Molina, S.; Botella, F. De-noising of short-period seismograms by wavelet packet transform. *Bull. Seismol. Soc. Am.* **2003**, *93*, 2554–2562. [CrossRef]
29. Botella, F.; Rosa-Herranz, J.; Giner, J.J.; Molina, S.; Galiana-Merino, J.J. A realtime earthquake detector with prefiltering by wavelets. *Comput. Geosci.* **2003**, *29*, 911–919. [CrossRef]
30. Zhang, H.; Thurber, C.; Rowe, C. Automatic P-wave arrival detection and picking with multiscale wavelet analysis for single-component recordings. *Bull. Seismol. Soc. Am.* **2003**, *93*, 1904–1912. [CrossRef]
31. Parolai, S.; Galiana-Merino, J.J. Effect of transient seismic noise on estimates of H/V spectral ratios. *Bull. Seismol. Soc. Am.* **2006**, *96*, 228–236. [CrossRef]
32. Hafez, A.G.; Khan, M.T.; Kohda, T. Clear P-wave arrival of weak events and automatic onset determination using wavelet filter banks. *Digit. Signal Process.* **2010**, *20*, 715–723. [CrossRef]
33. Beenamol, M.; Prabavathy, S.; Mohanalin, J. Wavelet based seismic signal denoising using Shannon and Tsallis entropy. *Comput. Math. Appl.* **2012**, *64*, 3580–3593. [CrossRef]
34. Galiana-Merino, J.J.; Rosa-Herranz, J.L.; Rosa-Cintas, S.; Martinez-Espla, J.J. SeismicWaveTool: Continuous and discrete wavelet analysis and filtering for multichannel seismic data. *Comput. Phys. Commun.* **2013**, *184*, 162–171. [CrossRef]
35. Gaci, S. The use of wavelet-based denoising techniques to enhance the first-arrival picking on seismic traces. *IEEE Trans. Geosci. Remote. Sens.* **2013**, *52*, 4558–4563. [CrossRef]
36. Shang, X.; Li, X.; Weng, L. Enhancing seismic P phase arrival picking based on wavelet denoising and kurtosis picker. *J. Seismol.* **2018**, *22*, 21–33. [CrossRef]
37. Tukey, J.W. *Exploratory Data Analysis*; Addison-Wesley: Boston, MA, USA, 1997; ISBN 978-0-201-07616-5. OCLC 3058187.
38. Donoho, D. De-noising by soft-thresholding. *IEEE Trans Inf Theory* **1995**, *41*, 613–627. [CrossRef]
39. Stein, C.M. Estimation of the mean of a multivariate normal distribution. *Ann Stat.* **1981**, *9*, 1135–1151. [CrossRef]
40. To, A.C.; Moore, J.R.; Glaser, S.D. Wavelet denoising techniques with applications to experimental geophysical data. *Signal Process.* **2009**, *89*, 144–160. [CrossRef]

41. Valencia, D.; Orejuela, D.; Salazar, J.; Valencia, J. Comparison Analysis Between Rigrsure, Sqtwolog, Heursure and Minimaxi Techniques Using Hard and Soft Thresholding Methods. In *Proceedings of the 2016 XXI Symposium on Signal Processing, Images and Artificial Vision (STSIVA), Bucaramanga, Colombia, 31 August–2 September 2016*; IEEE Xplore: New York, NY, USA, 2016; ISBN 978-1-5090-3798-8.
42. Zhao, R.M.; Cui, H.M. Improved Threshold Denoising Method Based on Wavelet Transform. In Proceedings of the 7th International Conference on Modelling, Identification and Control (ICMIC 2015), Sousse, Tunisia, 18–20 December 2015.

minerals

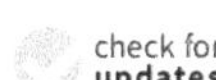

Article

A Numerical Study of Separation Performance of Vibrating Flip-Flow Screens for Cohesive Particles

Chi Yu, Runhui Geng and Xinwen Wang *

School of Chemical and Environmental Engineering, China University of Mining and Technology (Beijing), Beijing 100083, China; bqt1800301011@student.cumtb.edu.cn (C.Y.); sqt1900301005@student.cumtb.edu.cn (R.G.)
* Correspondence: xinwen.w@cumtb.edu.cn

Abstract: Vibrating flip-flow screens (VFFS) are widely used to separate high-viscosity and fine materials. The most remarkable characteristic is that the vibration intensity of the screen frame is only 2–3 g (g represents the gravitational acceleration), while the vibration intensity of the screen surface can reach 30–50 g. This effectively solves the problem of the blocking screen aperture in the screening process of moist particles. In this paper, the approximate state of motion of the sieve mat is realized by setting the discrete rigid motion at multiple points on the elastic sieve mat of the VFFS. The effects of surface energy levels between particles separated via screening performance were compared and analyzed. The results show that the flow characteristics of particles have a great influence on the separation performance. For 8 mm particle screening, the particle's velocity dominates its movement and screening behavior in the range of 0–8 J/m^2 surface energy. In the feeding end region (Sections 1 and 2), with the increase in the surface energy, the particle's velocity decreases, and the contact time between the particles and the screen surface increases, and so the passage increases. When the surface energy level continues to increase, the particles agglomerate together due to the effect of the cohesive force, and the effect of the particle's agglomeration is greater than the particle velocity. Due to the agglomeration of particles, the difficulty of particles passing through the screen increases, and the yields of various size fractions in the feeding end decrease to some extent. In the transporting process, the agglomerated particles need to travel a certain distance before depolymerization, and the stronger the adhesive force between particles, the larger the depolymerization distance. Therefore, for the case of higher surface energy, the screening percentage near the discharging end (Sections 3 and 4) is greater. The above research is helpful to better understand and optimize the screening process of VFFS.

Keywords: vibrating flip-flow screen; DEM; wet stick material; JKR model; separation performance

Citation: Yu, C.; Geng, R.; Wang, X. A Numerical Study of Separation Performance of Vibrating Flip-Flow Screens for Cohesive Particles. *Minerals* **2021**, *11*, 631. https://doi.org/10.3390/min11060631

Academic Editors: Daniel Saramak, Marek Pawełczyk and Tomasz Niedoba

Received: 14 May 2021
Accepted: 8 June 2021
Published: 14 June 2021

1. Introduction

Flip-flow screening technology is a new concept of screening technology that has been widely used and promoted in recent years. The VFFS has a wide range of applications in many fields, such as the fine coal screening process, cyclic screening of ore grinding products by high-pressure roller mill, and resource utilization of building solid waste [1,2]. Compared to traditional vibrating screens, such as linear vibrating screens and circular vibrating screens, the VFFS has the following advantage: small vibration intensity of main screen frame (2–3 g), therefore the dynamic load on the foundation is small; high vibration intensity of the sieve mat (up to 30–50 g). Furthermore, the VFFS is extremely friendly to the screening of viscous and wet fine-grained material, and it is not easy to block apertures on the screen surface while ensuring high screening efficiency and processing capacity. Due to the existence of water content between viscous and wet particles, there is a liquid bridge force between particles; particles will gather into clusters when the cohesion between particles is strong enough. When using traditional vibrating screens to process the wet and fine particles, the vibration intensity is not enough to make the agglomerated particles depolymerized, and the screens are extremely prone to blockage, adhesion, and compaction, which deteriorates the screening process [3]. The vibration frequency of VFFS is generally lower than a traditional screen, but through the large deformation of the elastic sieve mat, the peak acceleration is easy to produce. The vibration response of the sieve mat agitates the particle bed to deagglomerate the agglomerated particles. This drives the fine particles to flow down the bed and then pass through the screen to become the undersized product. The elastic sieve mat agitates the bed to depolymerize the agglomerated particles.

Standish constructed a single-particle model to investigate particle motion base on the reaction kinetics and probability theory. However, the collision between particles is not considered [4,5]. Soldinger developed a semi-mechanical phenomenological model of a linear vibrating screen, taking into account the stratification and passage [6]. Soldinger further extended the model after considering the material loading effect and the screening efficiency of different size particles [7]. The actual screening process is very complicated, and particle movement is affected by many conditions. At present, the discrete element method (DEM) simulation is an effective method for the simulation of granular systems, which has been used in various industrial processes. Cleary et al. quantitatively investigated the particle flow and screening performance of an industrial double-deck banana screen with different accelerations based on DEM simulation [8,9]. Davoodi et al. reported the effect of the aperture shape and the material on the particle flow and sieving performance [10]. Dong et al. simulated the screening process with the discrete element method and studied the influence of rectangular aperture shapes, with different aspect ratios, on material movement and screening efficiency [11]. Zhao et al. studied the influence of the motion parameters of the linear and circular vibration screens on the screening performance [12]. Wang et al. used the discrete element and the finite element methods to study the influence of vibration parameters on the screening efficiency of the vibrating screen. In addition, the distribution of stress and deformation on the screen surface under different vibration conditions has also been reported [13].

The above studies are mostly focusing on dry particulate systems, which are based on the Hertz–Mindlin model. In the actual screening process, due to the small particle size, large specific surface area, and external moisture, the fine particles easily agglomerate with each other to form large-size particles. The particles agglomerate together and move as a whole, making the screening process difficult. Limtrakul et al. reported that fine particles in a fluidized bed have particle agglomeration and stagnation regions due to high cohesion and confirmed the influence of vibration on improving fluidization through experiments [14]. Yang et al. investigated the influence of surface energy on the transition behavior of Geldart A-type particles from a fixed bed to a bubbling bed through a two-dimensional DEM-CFD simulation [15]. Cleary et al. reported the effect of cohesion between particles on particle flow over a double-deck banana screen [16]. At present, there are few numerical simulation studies on the movement and separation of viscous and wet material on VFFS.

In this study, the elastic sieve mat of the VFFS is discretized into multiple units by testing the movement of each unit body. According to the phase relationship of the unit body, it can describe the kinematics of the entire elastic sieve mat. The motion of each point on the sieve mat can be transformed into a function form by the Fourier series, which is used as the basis for setting the motion of the VFFS model. The effects of different adhesion levels on particle flow and screening performance on VFFS were compared and analyzed, which is helpful to better understand and optimize the screening process of the VFFS.

2. Simulation Methods

2.1. Contact Model of Particles

Due to clay and water present on the particle surface, there is a cohesive force between particles. The commonly used Hertz–Mindlin contact model struggles to comprehensively analyze the mechanical behavior between wet particles and between particles and the screen surface. The Hertz–Mindlin with JKR contact model, which considers the cohesive force, can better simulate the behavior of viscous and wet particles. Taking into account the effect of the surface energy (adhesion force) between the particles on the movement and screen penetration, the calculation of the normal elastic contact force is based on the Johnson–Kendall–Roberts theory [17,18].

Figure 1 shows the contact process of two cohesive particles. R_1 and R_2 represent the radius of Particle 1 and 2, respectively (mm). a stands for the contact radius between the particles (mm), and a_0 is the radius of the contact surface considering the adhesion (mm). δ_n is the amount of normal overlap (mm). Due to the cohesive force on the contact surface, the contact radius of these two particles extends from a to a_0.

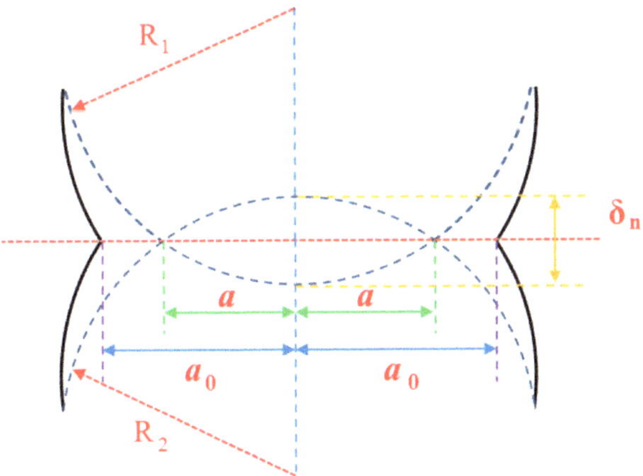

Figure 1. Deformation surface (rough line) considering cohesive force.

The cohesive force between the wet and viscous particles is set as $W(\text{J}/\text{m}^2)$, which can be obtained by Equation (1).

$$W = \gamma_1 + \gamma_2 + \gamma_{12} \tag{1}$$

where γ_1 is the surface energy of Particle 1 (J/m^2); γ_2 is the surface energy of Particle 2 (J/m^2); γ_{12} stands for the interface energy between Particles 1 and 2 (J/m^2). When the material of the particles is the same, the interface energy is 0 J/m^2, that is, $\gamma_{12} = 0$, $\gamma_1 = \gamma_2 = \gamma$, therefore, $W = 2\gamma$.

$$a = \sqrt{\delta_n R^*} \tag{2}$$

$$\delta_n = \frac{a_0^2}{R^*} - \sqrt{\frac{4\pi\gamma a_0}{E^*}} \tag{3}$$

$$\frac{1}{R^*} = \frac{1}{R_1} + \frac{1}{R_2} \tag{4}$$

$$\frac{1}{E^*} = \frac{1 - v_1^2}{E_1} + \frac{1 - v_2^2}{E_2} \tag{5}$$

Here γ is the surface energy between wet particles (J/m^2); R^* is the equivalent contact radius (mm); E^* is the equivalent elastic modulus (N/m^2); E_1, E_2 represent the elastic modulus of Particle 1 and 2, respectively (N/m^2); v_1 v_2 are the Poisson's ratio of these two particles, respectively (-).

Then, the normal elastic contact force $F_{JKR}(\text{N})$ between the wet particles can be calculated by Equation (6):

$$F_{JKR} = -2\sqrt{2\pi W E^* a_0^3} + \frac{4E^* a_0^3}{3R^*} \tag{6}$$

When the surface energy of the viscous particle is 0 J/m^2, the model F_{JKR} is simplified to the contact force F_{Hertz}.

2.2. The DEM Model Setting of VFFS

The structures of the VFFS and elastic sieve mat are presented in Figure 2. Different from the traditional vibrating screens, the VFFS consists of two vibrating frames, including the main screen frame and the floating screen frame. The beams of the two frames are arranged in a staggered layout. When the exciter mounted on the main screen frame is operated, both the screen frames move relative to each other through the effect of rubber shear springs. The elastic sieve mats are periodically stretched and slackened to generate peak acceleration, typically 30–50 times gravity.

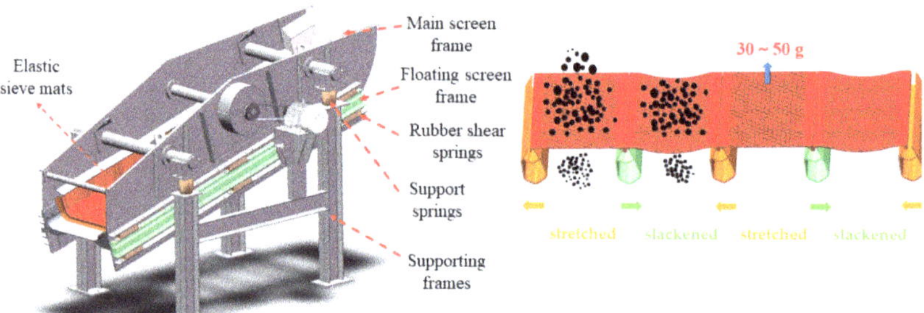

Figure 2. Structures of the VFFS and elastic sieve mat.

For traditional screening equipment such as circular vibrating screens and linear vibrating screens, the vibration parameters of the screen surface are consistent with the vibration response of the screen frame. Therefore, it is relatively easy to set the model of the traditional vibrating screen in the discrete element simulation. Many scholars have already done many in-depth studies in these fields [19–21]. For the VFFS, the vibration response of each position on the elastic sieve mat is different. The accelerometer is used to test the amplitude response at different positions on the elastic sieve mat. Figure 3 shows the measuring displacement of the midpoint of the sieve mat.

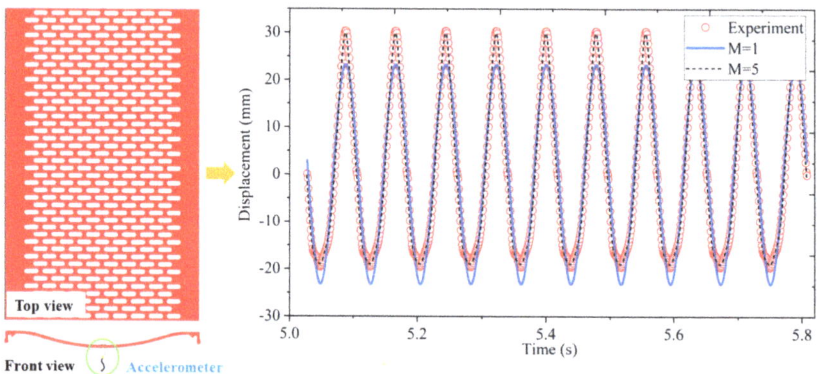

Figure 3. Fourier series analysis of amplitude at the midpoint of the sieve mat.

The displacement signal of the measuring point on the screen surface is not a regular simple harmonic function but periodic. Fortunately, any periodic function of time can be represented by the Fourier series as an infinite sum of sine and cosine terms [22]. Its Fourier series representation is given by Equation (7).

$$(t) = \alpha + \sum_{n=1}^{\infty} [a_n cos(n\omega t) + b_n sin(n\omega t)] \tag{7}$$

$$\alpha = \frac{2}{\tau} \int_0^\tau x(t)dt$$

$$a_n = \frac{2}{\tau} \int_0^\tau x(t)cos(n\omega t)dt$$

$$b_n = \frac{2}{\tau} \int_0^\tau x(t)sin(n\omega t)dt$$

where $\omega = 2\pi/\tau$ is called the fundamental frequency (rad/s) and α, a_1, a_2, ... , b_1, b_2, ... are constant coefficients (-). The M (intercepted order of Fourier series) has a direct impact on the accuracy of the calculation results. The larger the M, the closer the analysis result is to the accurate value [23], but it will also affect the solution efficiency. Then, we take the amplitude of the midpoint as an example for the Fourier analysis. Within a motion cycle, the peak value of the amplitude is 30.27 mm. In contrast to the signals analyzed by the Fourier series with the measured values, the results are shown in Figure 3. The mean square error (MSE) of a period and the relative error (RE) of maximum amplitude in the time domain are used to evaluate the change between the measured amplitude and the Fourier series analysis result. The results are shown in Table 1.

Table 1. Measured amplitude and analyzed amplitude by Fourier series on the midpoint.

Intercepted Order M	M = 1	M = 2	M = 3	M = 4	M = 5
Maximum amplitude (mm)	23.23	28.10	28.79	29.37	29.88
RE (%)	17.19	2.63	2.0	1.57	1.10
MSE	13.06	0.99	0.59	0.40	0.20

When the M is equal to one, the MSE is 13.06. When the M is equal to five, the MSE reduces to 0.2. Meanwhile, the RE is only 1.1%. Therefore, in this paper, the intercepted order of all amplitudes analyzed by the Fourier series is taken as five. The testing vibration amplitudes of each point on the elastic sieve mat are shown in Figure 4. It can be seen that the vibration amplitudes on the elastic sieve mat are symmetrically distributed, the midpoint has a large amplitude, and the edge measuring point has a relatively small amplitude. The movement of each point can be transformed into a function by the above-mentioned Fourier analysis method.

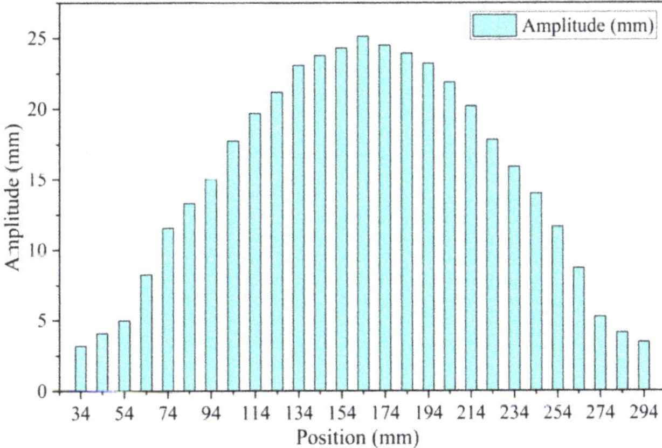

Figure 4. Vibration amplitude of each measuring point of the elastic sieve mat.

Further, the sieve mat is discretized into multiple units, and the simulation of the approximate continuous flexible motion is realized through the setting of multi-point rigid motion, as shown in Figure 5.

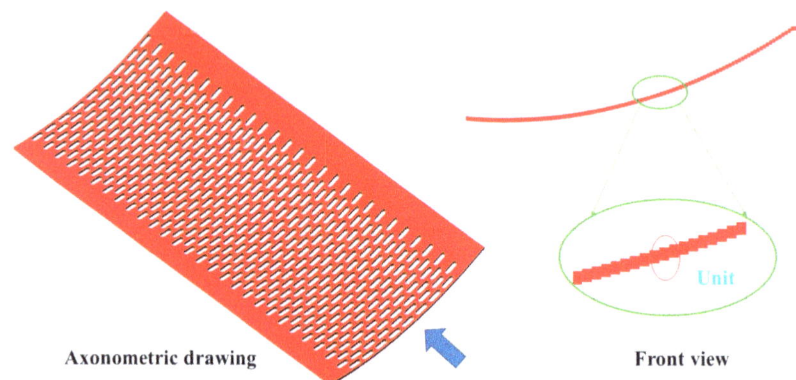

Figure 5. DEM model of the elastic sieve mat.

2.3. Simulation Conditions

Figure 6a shows the DEM modeling schematic of the VFFS system. The VFFS used in the simulation process is specifically composed of eight elastic sieve mats, each with a size of 328 mm × 650 mm. The screen aperture is 8 mm × 25 mm, and the inclination angle of the screen is 15°. In the simulation, the undersized product is divided into four parts equally by using 633.6 mm as the length interval unit, namely Sections 1–4. Section 5 is used to collect the oversized product. The feeding system is composed of a silo and a vibrating feeder. The material properties are shown in Figure 6b, and the simulation parameters in the DEM are shown in Table 2 [24,25]. It is worth noting that the impact of particle shape on the screening process is not considered in this study. The particles used in this simulation are all homogeneous spherical particles.

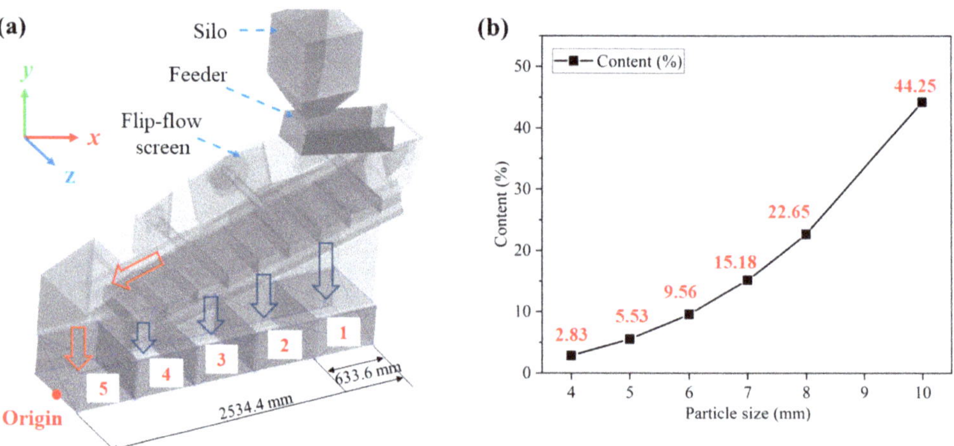

Figure 6. (**a**) Schematic of the DEM model of the VFFS system; (**b**) material properties of the sample.

Table 2. Modeling condition in EDEM.

Material Property	Poisson's Ratio (-)	Shear Modulus (Pa)	Density (kg/m^3)
Particle	0.250	2.200×10^8	2456
Polyrethane	0.499	1.157×10^6	1200
Steel	0.300	7.692×10^{10}	7850
Collision property	Coefficient of restitution	Coefficient of static friction	Coefficient of rolling friction
Particle-particle	0.50	0.154	0.10
Particle-polyrethane	0.25	0.500	0.01
Particle-steel	0.30	0.154	0.01
VFFS parameters			
Vibration parameter	The vibration frequency of 776 r/min, screen inclination of 15°		
Screen parameters	Screen length and width with 2624 and 650 mm, respectively		
Material properties	The total mass of 5.81 kg		

3. Effect of Surface Energy Level on Separation Performance

During the screening process, there are always some fine particles existing in the oversized products and some coarse particles in the undersized products. The screening efficiency and total misplaced material were used to assess the screening performance in this paper. The calculation formulas are as follows [26,27]:

$$\eta = E_c + E_f - 100$$
$$E_c = \frac{\gamma_o \times O_c}{F_c^r} \times 100 \tag{8}$$
$$E_f = \frac{F_f^r - \gamma_o \times O_f}{F_f^r} \times 100$$

$$M_o = M_c + M_f$$
$$M_c = 100 \times \gamma_u U_c \tag{9}$$
$$M_f = 100 \times \gamma_o O_f$$

where the η is the screening efficiency (%), E_c and E_f stand for the effective placement efficiency of the coarse particles (%) and the effective placement efficiency of fine particles (%), respectively. The M_o is the total misplaced material (%), M_c and M_f are the misplaced material of coarse particles (%) and the misplaced material of fine particles (%), respectively. The γ_o represents the yield of oversized product (%), γ_u is the yield of undersized product (%), O_f is the ratio of fine particles in the oversized product (%), O_c is the ratio of coarse particles in the oversized product (%), F_c^r is the ratio of coarse particles in the feeding (%), and F_f^r is the ratio of fine particles in the feeding (%).

Figure 7 shows the flow characteristics of material on VFFS with three surface energy levels (4, 20, and 36 J/m^2). In the case of the surface energy of 4 J/m^2, when the particles enter the screen, the vibration of the sieve mat quickly enables the material to spread on the screen surface. A larger amount of material pass through the screen in Section 1. As the screening process progresses along the direction of material flow, the amount of penetration in other sections gradually decreases. When the surface energy is 20 J/m^2, compared to the case of 4 J/m^2, the yield of material in Section 1 is reduced. This section mainly promotes the depolymerization of agglomerated particles. Meanwhile, the yield of material in Section 2 is increased. When the surface energy is 36 J/m^2, there is a great cohesion force between the particles, and the agglomerated particles need a longer movement distance to complete the depolymerization process. In Sections 1 and 2, which near the feeding end, the yield of the undersized product is low, and more particles are concentrated in Sections 3 and 4, near the discharging end. To further deepen the understanding of the screening process of VFFS, quantitative analysis was carried out on the products of each section.

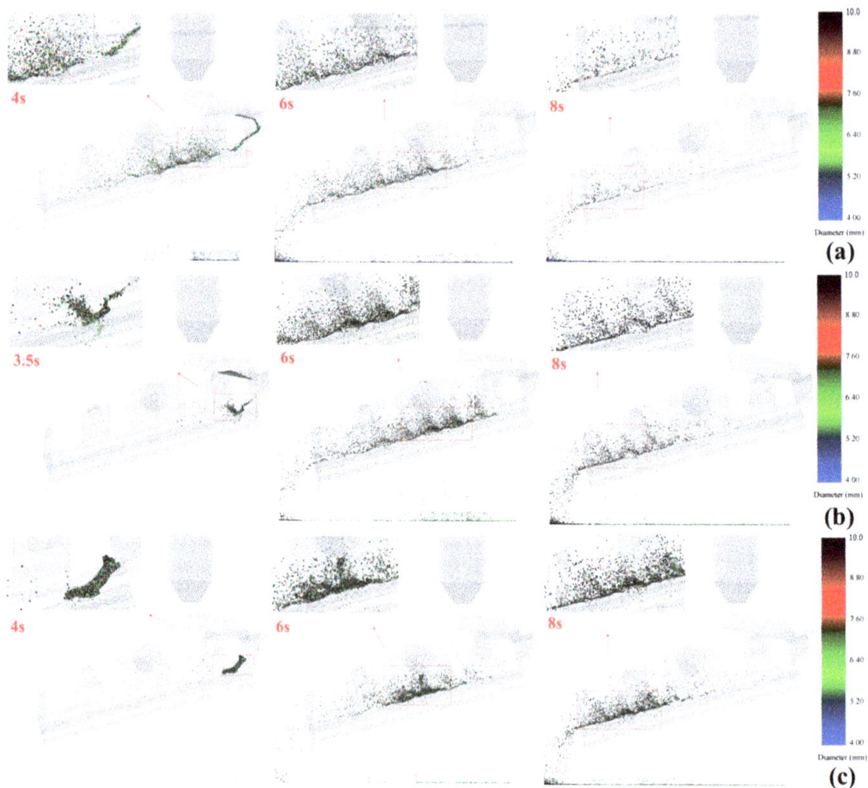

Figure 7. The flow behavior of material on VFFS with different surface energy levels. (**a**) 4, (**b**) 20, and (**c**) 36 J/m^2.

3.1. The Yield of Each Section of VFFS

Figure 8a shows the distribution of product yields between undersized and oversized products with different surface energy levels, where the distribution is related to the surface energy between particles. With the increase in the surface energy level between particles, the yield of the undersized product first increased and then decreased, and at the same time, the yield of oversized products decreased and then increased. This means that for each surface energy level, the sum of the undersized and oversized product is 100%. In the case of the surface energy level of 0–8 J/m^2, the particle movement speed dominates the movement and separation behavior of the particles. With the increase in the surface energy level within the range of 0–8 J/m^2, the particle movement speed decreases, increasing the contact time between the particle and the screen surface. Therefore, the amount of material passing through the screen increases. When the cohesive force continues to increase to a certain level, the particles agglomerate together, and the impact of particle agglomeration is greater than the particle movement speed. As more fine particles agglomerate together, their size increases to greater than the aperture size, and the material screening percentages decreases. The stronger the surface energy between the particles, the longer the distance the agglomerated particles need to deagglomerate. More fine particles finally enter the oversized products, so the yield of undersized products drops again. When the surface energy level is 36 J/m^2, around 70% of the material enters the oversized product.

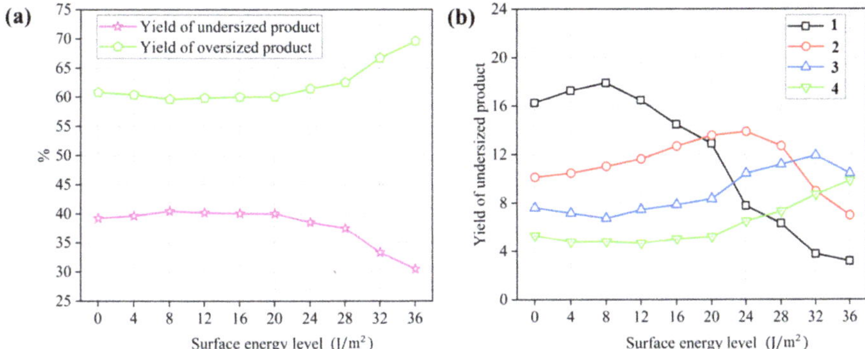

Figure 8. (**a**) Distribution of the yields between the undersized and oversized product with different surface energy levels; (**b**) The undersized product yields of each section with different surface energy levels.

The yield of each section of the undersized product is further analyzed, and the results are shown in Figure 8b. It can be seen that with the increase of the surface energy level, the yield of the undersized product in Section 1 shows a trend of increasing and then decreasing. When the surface energy is 8 J/m^2, the yield in Section 1 is the largest, which is 17.91%. When the surface energy of particles continues to increase, the yield of Section 1 will gradually decrease. For the yield of Section 2, with the increase of the surface energy, it also shows the law of first increasing and then decreasing. However, the maximum yield appears at the surface energy of 24 J/m^2. With the increase of the surface energy, the yield in Section 3 first decreases and then increases to a maximum value when the surface energy is 32 J/m^2 and then decreases again. The yield in Section 4 first decreases and then increases. For the surface energy levels of 0, 4, 8, 12 and 16 J/m^2, the undersized product yield of each section gradually decreases along the direction of the material flow, and Section 1 accounts for the largest proportion. When the surface energy is 20 J/m^2, the yield in Section 2 is greater than the yield of Section 1. When the surface energy level continues to increase to 32 J/m^2, the yield of Section 3 is greater than the products of Sections 1 and 2. In Section 1, less than 4% of the particles pass through the screen. As the surface energy increases, the section with the maximum yield moves toward the discharging end. That is, the higher the adhesion, the longer the distance required for the depolymerization of the agglomerated particles.

3.2. The Yield Accounted for Size Fraction in Different Sections

Figure 9 shows the yield accounted for different size fractions in different sections, during the screening process of the VFFS. For 8 mm particle screening and different surface energy levels, the 10 mm particles are the oversized product and are all concentrated in Section 5. When the surface energy is 0, 4, and 8 J/m^2, the 4 and 5 mm particles are mainly concentrated in Section 1, accounting for about 50% of this size fraction. With the increase of particle size, the yield accounts for this size fraction in Section 1 gradually decreases. For the 8 mm particles, the yields of this size fraction are 16.59%, 18.41%, and 19.80%, respectively. In addition, for the case of these surface energy levels, as the surface energy between particles increases, the yield of each size fraction also increases. This is because the increase in the level of adhesion reduces the speed of particle movement. The contact time between the particles and the screen surface is increased, increasing the yield of the particles of each size fraction. For the cases of the surface energy of 12, 16, and 20 J/m^2, the yield of each size fraction in Section 1 gradually decreased, and the yield of 4 and 5 mm particles decreased to 51.93%, 44.50%, 36.63% and 47.98%, 38.87%, 35.40%, respectively. For the case of the surface energy levels of 24, 28, and 32 J/m^2, in the product of Section 1, the yield accounted for the size fraction of each sized particle further decreases. Meanwhile, it is worth noting that the yield of 4 mm particles in Section 1 is slightly smaller than that of

5 mm particles, which is due to the surface energy of particles having a greater influence on fine ones.

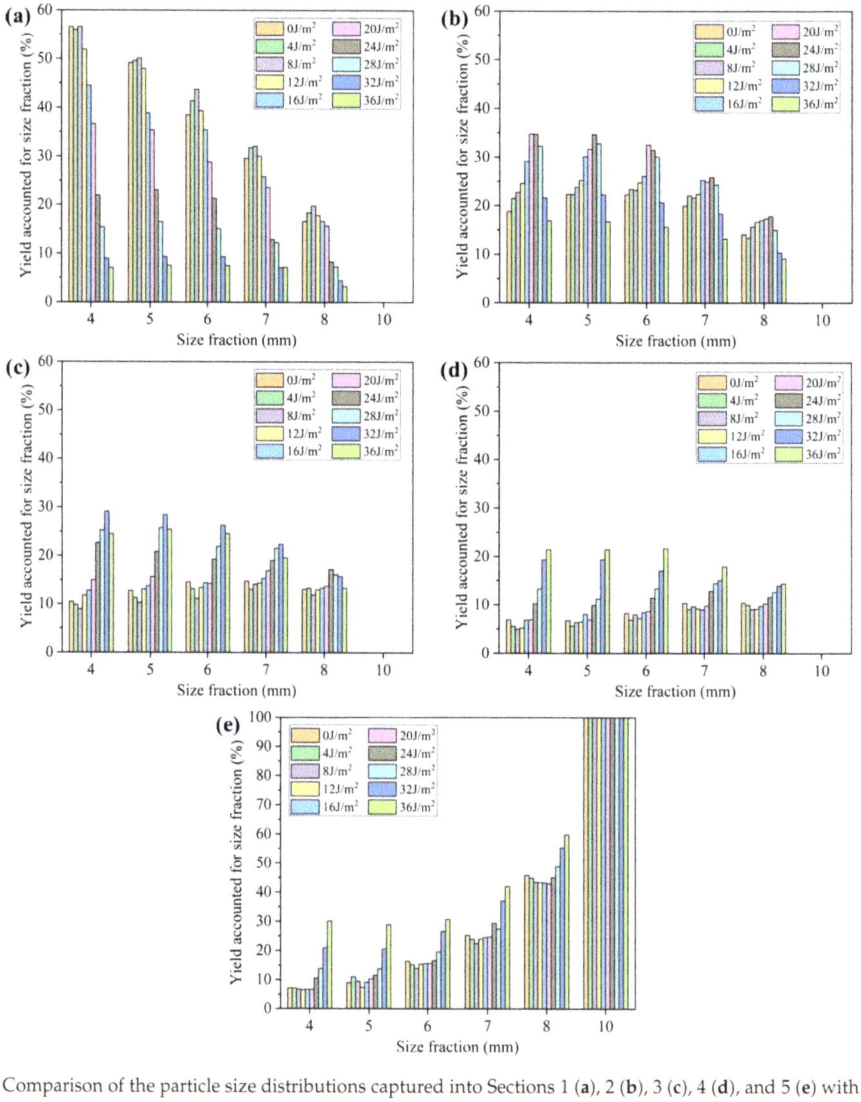

Figure 9. Comparison of the particle size distributions captured into Sections 1 (**a**), 2 (**b**), 3 (**c**), 4 (**d**), and 5 (**e**) with different surface energy levels.

The yield of each sized particle in Section 2 was observed. For the case of 0, 4, and 8 J/m^2 surface energy, the yield of the undersized product in Section 2 was significantly lower than Section 1. Within the range of 0–8 J/m^2, as the surface energy increased, the yield of fine particles increased, and this phenomenon was more obvious when the adhesion level was higher. When the surface energy was 24 J/m^2, the yields of each size fraction in Section 2 begin to exceed those in Section 1. The yields of large-sized materials in Sections 3 and 4 are generally higher than in small-sized materials. As the surface energy increases, more fine particles are deagglomerated under the movement of the sieve mat in Sections 3 and 4, and the yield of fine particles in the product begins to exceed that of coarse particles, becoming the dominant product in Sections 3 and 4. For particles with a higher level of adhesion, a longer movement distance, that is, a higher external energy supplement, is required to complete the depolymerization of agglomerated particles. The higher the adhesive force level, the closer the maximum yield section in the undersized product is to the discharging end. Moreover, the smaller the particle size, the more obvious this phenomenon.

3.3. The Screening Percentage of Different Size Fractions of Different Sections

The screening percentages of various size fractions in different Sections of VFFS with different surface energy levels are shown in Figure 10. In the products of Section 1, for the particles at the surface energy of 24, 28, 32 J/m^2, and 36 J/m^2, it can be seen that the screening percentages of various size fractions are significantly lower than that of other surface energy levels. Taking 4 mm particles as an example, the screening percentages of 4 mm particles are 21.93%, 15.41%, 9.00%, and 7.10%, respectively. The main effect of Section 1 is to promote the depolymerization of agglomerated particles. The longer the transporting distance of agglomerated particles, the better the depolymerization effect. It can be observed that for 4 mm particles under the case of the surface energy level of 20 J/m^2, the screening percentage in Section 1 is 36.63%, in Section 2 increases to 54.89%, and the screening percentages in Sections 3 and 4 are 52.45% and 50.74%, respectively. When the cohesive force level continues to increase to 28 J/m^2, the screening percentage in Section 1 is 15.41%, in Section 2 it is 38.08%, and increases to 48.23% and 48.89% in Sections 3 and 4, respectively. When the surface energy is 36 J/m^2, the screening percentage in Section 1 is only 7.10%, and further increases to 18.19%, 32.24%, and 41.55%, respectively. Compared with the coarse particles, the surface energy level has a more significant effect on fine particles. After the depolymerization of fine particles, the screening percentage of fine particles will be significantly improved.

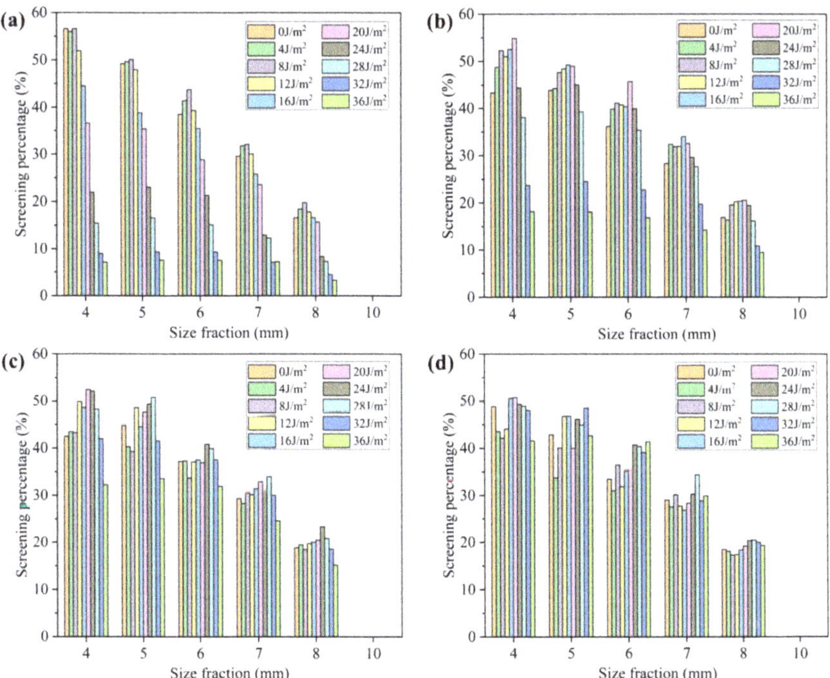

Figure 10. Comparison of the screening percentages of various size fractions in Sections 1 (**a**), 2 (**b**), 3 (**c**), and 4 (**d**) of VFFS with different surface energy levels.

3.4. The Screening Performance of Various Size Fractions in Different Sections and Screen Length

Figures 11 and 12 show the screening efficiency and misplaced material of various size fractions in different sections and screen lengths of the VFFS with different surface energy levels. The particles shape in the simulation are spherical, so in the actual simulation process, no coarse particles enter the undersized products. The effective placement efficiency of the coarse particles is 100%, and the misplaced material of coarse particles is 0%. Therefore, the screening efficiency is equal to the effective placement efficiency of fine particles, and the total misplaced material is equal to the misplaced material of fine particles. For 8 mm particle screening, when the surface energy between particles is $0\,\mathrm{J/m^2}$, the screening efficiency in Section 1 reaches the maximum of 29.15%, and the total misplaced material is 39.50%. With the flow of material, the screening efficiency in Sections 1–4 decreases gradually. When the surface energy increases to 5 and $8\,\mathrm{J/m^2}$, the screening efficiency increases in Sections 1 and 2, and the total misplaced material decreases, which is mainly due to the surface energy between particles reducing the movement speed of particles and increasing the residence time of particles on the screen surface, thus increasing the screening efficiency. After the surface energy of $16\,\mathrm{J/m^2}$, the influence of particle agglomeration begins to be greater than particle velocity, and the screening efficiency starts to decrease. The screening efficiency of Section 2 begins to be greater than that of Section 1. In addition, the screening efficiency of Sections 3 and 4 are higher than those of the levels 0, 5, 8, and $12\,\mathrm{J/m^2}$. For the case of $24\,\mathrm{J/m^2}$, the maximum screening efficiency appears in Section 3, which is 30.54%. When the surface energy increases to $32\,\mathrm{J/m^2}$, the screening efficiency of Section 4 is the highest, which is 27.88%. For different surface energy levels, the screening efficiency of particles increased with the increase in screening length. When the surface energy is $8\,\mathrm{J/m^2}$, the screening efficiency of VFFS is the highest, which is 72.52%, and the total misplaced material is 23.81%.

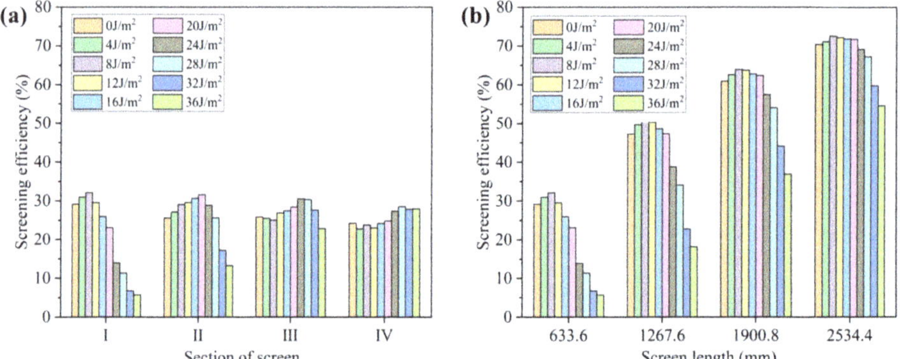

Figure 11. (**a**) Screening efficiency of different sections of the screen with different surface energy levels, (**b**) screening efficiency of different screen lengths with different surface energy levels.

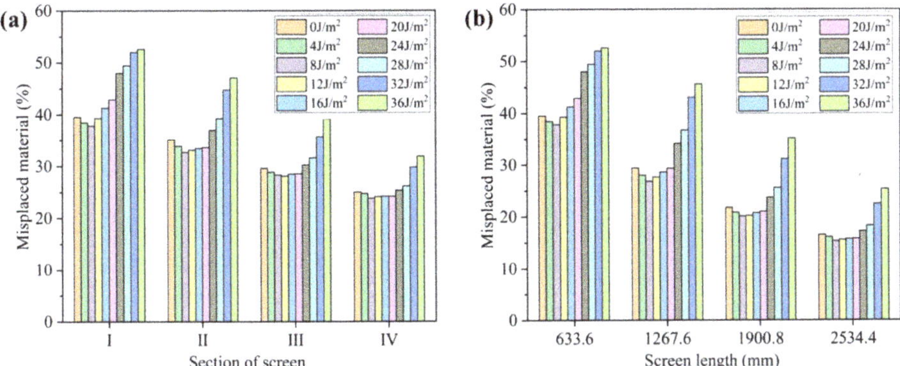

Figure 12. (**a**) Misplaced material of the different sections of the screen with different surface energy levels, (**b**) misplaced material of different screen lengths with different surface energy levels.

4. Conclusions

The following conclusions can be drawn from the above research.

(1) Due to the amplitude at each point on the sieve mat changing periodically, the motion can be transformed into a function form by the Fourier series. The DEM simulation of VFFS is realized by setting the multi-point rigid motion of the sieve mat.

(2) When the surface energy level is in the range of 0 to 8 J/m^2, the particle velocity in the feeding end region (Sections 1 and 2) dominates the movement behavior of particles passing through the screen. In Sections 1 and 2, the particle movement speed decreases, which increases the contact time between the particles and the screen surface, increasing the screening percentages. When the level of surface energy continues to increase, more fine particles are agglomerated together, which increases the screening difficulty. The effect of particle agglomeration in the feeding end is greater than its movement speed, and the screening percentages of each particle size in the feeding end have been reduced. Agglomerated particles need a certain transporting distance to deagglomerate. The stronger the surface energy between particles, the greater the distance the particles need to deagglomerate. Therefore, for the case of a higher surface energy level, close to the discharging end (Sections 3 and 4), the screening percentages of the material are greater.

(3) The screening efficiency increases with the increase in screen length for different surface energy levels. When the surface energy is 8 J/m^2, the screening performance of VFFS is better, with a screening efficiency of 72.52% and a total misplaced material of 23.81%.

Since the shape of the screen apertures of the elastic sieve mat is a straight slot, in the actual screening process, there is a situation that the strip particles pass through the screen. In future work, the influence of the shape characteristics of the particles on their movement and screening performance should be considered. Furthermore, we still need to carry out some full-scale screening experiments of wet particles based on the experimental VFFS.

Author Contributions: Conceptualization, C.Y. and X.W.; methodology, C.Y. and X.W.; software, C.Y.; validation, C.Y., R.G. and X.W.; formal analysis, C.Y.; investigation, C.Y.; resources, X.W.; data curation, C.Y.; writing—original draft preparation, C.Y. and R.G.; writing—review and editing, X.W.; visualization, X.W.; supervision, X.W.; project administration, X.W.; funding acquisition, X.W. All authors have read and agreed to the published version of the manuscript.

Funding: This research was funded by "the Fundamental Research Funds for the Central Universities" (No. 2020YJSHH17) (No. 2021YJSHH32).

Data Availability Statement: All data and models generated or used during the study appear in the submitted article.

Acknowledgments: The authors would like to thank the company of TianGong technology for its support in enabling this research.

Conflicts of Interest: The authors declare no conflict of interest.

References

1. Zhai, H.X. Determination of the operation range for flip-flow screen in industrial scale based on amplitude-frequency response. *J. China Coal Soc.* **2007**, *7*, 753–756.
2. Gong, S.; Oberst, S.; Wang, X. An experimentally validated rubber shear spring model for vibrating flip-flow screens. *Mech. Syst. Signal Process.* **2020**, *139*, 106619. [CrossRef]
3. Xiong, X.; Niu, L.; Gu, C.; Wang, Y. Vibration characteristics of an inclined flip-flow screen panel in banana flip-flow screens. *J. Sound Vib.* **2017**, *411*, 108–128. [CrossRef]
4. Standish, N. The kinetics of batch sieving. *Powder Technol.* **1985**, *41*, 57–67. [CrossRef]
5. Standish, N.; Bharadwaj, A.; Hariri-Akbari, G. A study of the effect of operating variables on the efficiency of a vibrating screen. *Powder Technol.* **1986**, *48*, 161–172. [CrossRef]
6. Soldinger, M. Interrelation of stratification and passage in the screening process. *Miner. Eng.* **1999**, *12*, 497–516. [CrossRef]
7. Soldinger, M. Influence of particle size and bed thickness on the screening process. *Miner. Eng.* **2000**, *13*, 297–312. [CrossRef]
8. Cleary, P.W.; Sinnott, M.D.; Morrison, R.D. Separation performance of double deck banana screens—Part 1: Flow and separation for different amplitudes. *Miner. Eng.* **2009**, *22*, 1218–1229. [CrossRef]
9. Cleary, P.; Sinnott, M.D.; Morrison, R.D. Separation performance of double deck banana screens—Part 2: Quantitative predictions. *Miner. Eng.* **2009**, *22*, 1230–1244. [CrossRef]
10. Davoodi, A.; Bengtsson, M.; Hulthén, E.; Evertsson, C. Effects of screen decks' aperture shapes and materials on screening efficiency. *Miner. Eng.* **2019**, *139*, 105699. [CrossRef]
11. Dong, K.; Esfandiary, A.H.; Yu, A. Discrete particle simulation of particle flow and separation on a vibrating screen: Effect of aperture shape. *Powder Technol.* **2017**, *314*, 195–202. [CrossRef]
12. Zhao, L.L.; Liu, C.S.; Yan, J.X.; Jiang, X.W.; Zhu, Y. Numerical simulation of particle segregation behavior in different vibration modes. *Acta Phys. Sin Ch. Ed.* **2010**, *59*, 2582–2588.
13. Wang, Z.; Liu, C.; Wu, J.; Jiang, H.; Zhao, Y. Impact of screening coals on screen surface and multi-index optimization for coal cleaning production. *J. Clean. Prod.* **2018**, *187*, 562–575. [CrossRef]
14. Limtrakul, S.; Rotjanavijit, W.; Vatanatham, T. Lagrangian modeling and simulation of effect of vibration on cohesive particle movement in a fluidized bed. *Chem. Eng. Sci.* **2007**, *62*, 232–245. [CrossRef]
15. Yang, F.; Thornton, C.; Seville, J. Effect of surface energy on the transition from fixed to bubbling gas-fluidised beds. *Chem. Eng. Sci.* **2013**, *90*, 119–129. [CrossRef]
16. Cleary, P.W.; Wilson, P.; Sinnott, M.D. Effect of particle cohesion on flow and separation in industrial vibrating screens. *Miner. Eng.* **2018**, *119*, 191–204. [CrossRef]
17. Cao, B.; Li, W.H.; Wang, N.; Bai, X.Y.; Wang, C.W. Calibration of Discrete Element Parameters of the Wet Barrel Finishing Abrasive Based on JKR Model. *Surf. Technol.* **2019**, *48*, 249–256.
18. Feng, X.; Liu, T.; Wang, L.; Yu, Y.; Zhang, S.; Song, L. Investigation on JKR surface energy of high-humidity maize grains. *Powder Technol.* **2021**, *382*, 406–419. [CrossRef]
19. Delaney, G.W.; Cleary, P.; Hilden, M.; Morrison, R.D. Testing the validity of the spherical DEM model in simulating real granular screening processes. *Chem. Eng. Sci.* **2012**, *68*, 215–226. [CrossRef]
20. Davoodi, A.; Asbjörnsson, G.; Hulthén, E.; Evertsson, M. Application of the Discrete Element Method to Study the Effects of Stream Characteristics on Screening Performance. *Minerals* **2019**, *9*, 788. [CrossRef]
21. Harzanagh, A.A.; Orhan, E.C.; Ergun, S.L. Discrete element modelling of vibrating screens. *Miner. Eng.* **2018**, *121*, 107–121. [CrossRef]
22. Singiresu, S.R. *Mechanical Vibrations*, 5th ed.; Prentice Hall: Upper Saddle, NJ, USA, 2011; pp. 72–80.
23. Zhang, Y.; Shi, D.; He, D.; Shao, D. Free Vibration Analysis of Laminated Composite Double-Plate Structure System with Elastic Constraints Based on Improved Fourier Series Method. *Shock. Vib.* **2021**, *2021*, 8811747. [CrossRef]
24. Li, Y.; Xu, Y.; Thornton, C. A comparison of discrete element simulations and experiments for 'sandpiles' composed of spherical particles. *Powder Technol.* **2005**, *160*, 219–228. [CrossRef]
25. Wu, B.; Zhang, X.; Niu, L.; Xiong, X.; Dong, Z.; Tang, J. Research on Sieving Performance of Flip-Flow Screen Using Two-Way Particles-Screen Panels Coupling Strategy. *IEEE Access* **2019**, *7*, 124461–124473. [CrossRef]
26. Jiang, H.; Wang, W.; Zhou, Z.; Jun, H.; Wen, P.; Zhao, Y.; Duan, C.; Zhao, L.; Luo, Z.; Liu, C. Simultaneous multiple parameter optimization of variable-amplitude equal-thickness elastic screening of moist coal. *Powder Technol.* **2019**, *346*, 217–227. [CrossRef]
27. Jiang, H.; Zhao, Y.; Duan, C.; Zhang, C.; Diao, H.; Wang, Z.; Fan, X. Properties of technological factors on screening performance of coal in an equal-thickness screen with variable amplitude. *Fuel* **2017**, *188*, 511–521. [CrossRef]

Article

Assessment of Selected Characteristics of Enrichment Products for Regular and Irregular Aggregates Beneficiation in Pulsating Jig

Tomasz Gawenda [1], Daniel Saramak [1], Agata Stempkowska [1,*] and Zdzisław Naziemiec [2]

1. Department of Environmental Engineering, Faculty of Civil Engineering and Resource Management, AGH University of Science and Technology, Mickiewicza 30 Av., 30-059 Cracow, Poland; gawenda@agh.edu.pl (T.G.); dsaramak@agh.edu.pl (D.S.)
2. Lukasiewicz-Institute of Ceramic Building Materials in Cracow, Cementowa 8 Str., 31-983 Cracow, Poland; z.naziemiec@icimb.pl
* Correspondence: stemp@agh.edu.pl

Abstract: Article concerns problem of jig beneficiation of mineral aggregates and focuses especially on problem of separation of hard-enrichable materials. Investigative programme covered tests in laboratory and semi-plant scale and material with different content of regular and irregular particles, along with various particle size fractions, was under analysis. Two patented solutions were utilized as methodological approach and densities and absorbabilities of individual products were determined and major novelty of approach consist in separate beneficiation of regular and irregular particles. Results of laboratory investigations showed that more favorable separation effectiveness was observed for the narrow particle size fractions of feed material. In terms of absorbability difference between separation products from I and IV layer was 0.4–0.5% higher for regular particles, and up to 0.5% higher for irregular grains. Differences in densities of respective products were 0.1% higher for regular particles. Results of semi-plant tests confirmed the outcomes achieved in laboratory scale. The qualitative characteristics of separation products in terms of micro-Deval and LA comminution resistance indices were one category higher for regular particles, and two categories higher for irregular grains, comparing to the raw material.

Keywords: aggregates; jig beneficiation; mineral processing; raw materials; separation

check for **updates**

Citation: Gawenda, T.; Saramak, D.; Stempkowska, A.; Naziemiec, Z. Assessment of Selected Characteristics of Enrichment Products for Regular and Irregular Aggregates Beneficiation in Pulsating Jig. *Minerals* **2021**, *11*, 777. https://doi.org/10.3390/min11070777

Academic Editor: Carlos Hoffmann Sampaio

Received: 27 May 2021
Accepted: 14 July 2021
Published: 17 July 2021

Publisher's Note: MDPI stays neutral with regard to jurisdictional claims in published maps and institutional affiliations.

1. Introduction

Jig beneficiation is a simple and economical method of raw materials enrichment. It is especially efficient for separation of minerals with relatively high differences of their densities. The environmental footprint of jig beneficiation is also low, especially in terms of dust and noise pollution but also due to relatively low water and energy consumption, comparing to flotational separation or chemical beneficiation, and also selected operations of mechanical enrichment [1–3]

An up to date review of the jigging operation fundamentals and outlines of directions for future research and developments were presented in [4]. The configuration, operational principles, and main applications of different jig types have been comprehensively reviewed. A description of the main theoretical approaches was also presented, with highlighting of strengths and weaknesses of operations. Gravity separation is, in general, quite simple method of raw material beneficiation and quite well documented in literature. This method is economically efficient and does not require an intensive consumption of energy and other media [5,6].

Jig beneficiation is most commonly used for coal preparation, where the difference in densities of coal and the waste rock are high [7,8]. In rock materials processing, the effectiveness of separation process might be lower, but it greatly depends on the degree of useful mineral liberation and the difference in densities of the material being separated. However it appears that apart from density several other features of the feed material are influential, like particle size and the shape of individual particles [9,10]. Results of various investigation show that after suitable preparation of the feed material i.e., separation of entire feed into narrow particle size fractions, together with distinction of regular and irregular particles, may significantly increase the effect of jig process beneficiation. The jig enrichment technology can also be used in the processing of other mineral resources, especially in removing of impurities at initial stages of aggregates, sand or gravel treatment [11]. It is also possible to recover in jigs building materials from demolition debris and road materials [12], electronic waste or plastic [13,14].

Industrial practice shows that jigs are common in plant operation, however not in all sectors of mining and mineral processing industry. Beneficiation of raw materials with relatively high differences of densities are not problematic, but in some aspects, especially when the difference between the density of useful mineral (i.e., than can be used effectively in further production) and the gangue, or between the two product to be separated, is lower, application of jigs is an issue [4].

There are relatively low number of publications concerning applications of jigging processes into aggregate enrichment. More results can be found in concrete separation especially the materials from demolition of building constructions [15,16]. Despite of that the problem of aggregate segregation is significant, because it influences improvement of qualitative characteristics of final products, through elimination of weaker and partially damaged particles, as well as weathered grains [17,18].

2. Materials and Methods

2.1. Research Significance

There has been done a significant development in the design of settling machines in recent years, resulted from a high demand for good quality aggregates [4,19,20]. An application of density separation methods into aggregate production can be used more effectively in industry due to the different lithological and physical properties of the separation products and variations in values of strength and absorbability parameters [21,22]. These applications are suitable for natural, recycled and anthropogenic aggregates [15,16] The article presents results of tests on a prototype and unique *SET* device (a separator for hard-enrichable particles), constructed especially for research purposes according to the patented concept (Patent PL 233318B1). The main and innovative idea presented in the paper concerns running the jig beneficiation process in more narrow particle size fraction, and separately for regular and irregular particles. Such an approach was not presented in literature.

2.2. Methodology

The enrichment process in a jig takes place as a result of the separation of the feed into fractions according to the selected physical feature, i.e., the density [23]. It can be effective in the air, water or other liquid medium lighter than the components of the enriched material. Rock materials with different densities can be characterized by different settling velocities when they are subjected to a vertical pulsating motion in a specific medium (i.e., water). The theoretical bases of separation in jigs have been described by Newton-Rittinger law. According to this law, the description of particle movement in a liquid under the influence of pulsation takes into account the limiting settling velocity in a situation in which the geometric sum of gravity and hydrostatic buoyancy forces, as well as the medium resistance that act on a single particle, equals zero. It is assumed that when the material layer in the jig loosens, each particle moves freely, regardless the presence of surrounding particles. The free settling ratio (Equation (1)) can be helpful in assessing the separability of the material under treatment.

$$e = \frac{d_1}{d_2} = \frac{\rho_2 - \rho_o}{\rho_1 - \rho_o} \tag{1}$$

where: index 1—particles of lower density; index 2—particles of higher density; d—particle diameter, [mm]; e—free settling ratio; ρ_1, ρ_2,—particle densities, [g/cm^3]; ρ_0—liquid density, [g/cm^3].

The free settling ratio, e, gives an indication of the scope of the feed material preparation needed for enrichment in jigs and defines the following natural limitations for the process, which, if not taken into account, can significantly reduce the efficiency of separation:

- feed should contain a minimum of equally settling particles i.e., differentiated in terms of densimetric and granulometric properties,
- the material must be homogeneous in terms of density distribution,
- the feed material should be proceeded into narrow granular classes.

Two most common efficiency indicators used in assessment of jigs performance are the probable error (Ep) and imperfection (I) [24]. The regularity and irregularity of particles can be measured by means of the Schultz caliper or on a bar slotted screen. The obtained values of shape and flatness coefficients indicate whether the particle can be regarded as a regularly shaped or not. The general principle is that if the shortest dimension of a particle is more than two time smaller than its longest dimension, the particle is regarded as irregular in shape. Also, particles that are porous have a natural ability to absorb a significant amount of water. Porosity was determined indirectly by determining the volume of water retained within an individual particle. The effectiveness of jig beneficiation was also analyzed in terms of a particle's water absorption properties or 'absorbability'—an index determined by measuring the amount of water absorbed by the particle immersed in water.

Primary aim of application of jigs in raw materials enrichment is a feed separation into products according to density of individual particles, but the material can be also separated according to the shape. Particle settling velocity is determined by density, size and shape of a particle and can be calculated on the bases of Formula (2), derived from heuristic considerations [25]:

$$v = 5.33 \sqrt{x} \sqrt{d_p} \sqrt{\left(\frac{k_1}{k_2}\right)} \tag{2}$$

where: $x = (\rho - \rho_0)/\rho$—density of reduced particle, ρ—density of particle, ρ_0—density of the liquid, d_p—projection diameter of particle, k_1—volumetric shape coefficient, k_2—dynamic shape coefficient.

On the basis of work [18,26] there can be noticed differences in settling velocities of regular and irregular particles (Table 1), while the greatest variations can be observed in upper product (layer IV).

Table 1. Settling velocities for regular and irregular particles in individual layers (jig products).

Number of Layer (Product) in the Jig	Settling Velocity, [m/s]		
	For Regular Particles	For Irregular Particles	Difference
I	0.21	0.18	0.03
II	0.20	0.16	0.04
III	0.19	0.14	0.05
IV	0.22	0.16	0.06

It can be also noticed that regular particles have higher values of settling velocities, due to lower resistance of their round edges. It is then possible to obtain a more effective separation process of regular and irregular particles in enrichment of a materials with narrow particle size fraction. This has been investigated in a semi-plant tests within the paper. The entire research program includes laboratory and semi-plant tests and in semi-plant scale the jig constructed especially for purposes of investigations, according to the concept of a patented invention (PL 233318B1), was used. Variables in laboratory tests were selected considering qualitative criteria (content of regular or irregular particles), and technological aspects of aggregate production (fine or coarse products). These variables entirely determine value of obtained products, thus control of these parameters may influence the economic effectiveness of a mineral aggregate production plant. Different levels of variables in experiments, in turn, were adopted on the bases of characteristics of typical commercial aggregate products.

3. Experimental

3.1. Characteristics of Testing Device

An innovative approach presented in the paper consists in a separate enrichment of regular and irregular particles. Narrow particle size fractions were divided from the material prior to the jigging process. The reason of applying the above procedure was that properties of particles in specific size fractions are more homogenous, and variations in values of individual features (i.e., density, absorbability, porosity) could be more visible. In the first stage of investigations laboratory scale experiments were performed, and then semi-plant tests were carried out. Preparation of feed for laboratory tests was possible thanks to the application of a patented solution (Figure 1) for production of regular and irregular aggregates (Patent PL233689). The solution has not been used in aggregate production so far, and the findings of the paper show that its application is justified and may give measurable effects.

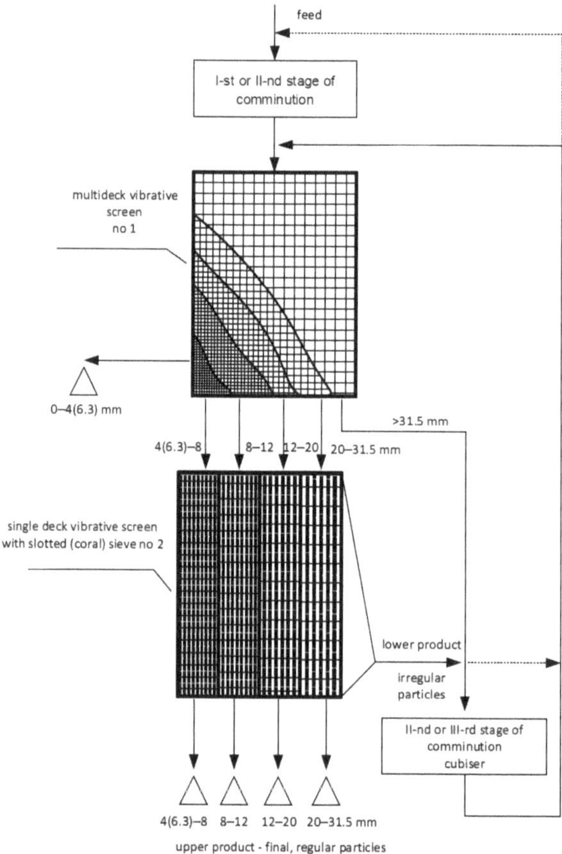

Figure 1. Idea of the aggregate production circuit with a closed recirculation for selective screening and crushing operations (Patent PL233689).

The circuit designed according to the innovative idea, can produce a final aggregate product with irregular particle contents as low as 2–3%. It requires only the application of quadratic and slotted mesh sieves, cooperating with the crusher working in a closed circuit, either operating on a first or second stage. The obtained irregular particles can be comminuted in the same crusher or in a secondary stage of crushing, for example in an impact crusher (cubiser) what can additionally improve the quality of the product. The contents of irregular particles in final products depends on capacity of the screen with slotted sieve and especially on the relation between the narrow particle fraction range and size of the slot in sieve. This sieve should be selected according to "$d_{max}/2$" principle, i.e., half of the maximum size of the particle fraction. For the reason that content of irregular particles is lower for coarser particle fractions and the screening efficiency for coarser particles is better, screening of irregular particles in coarser fractions will be easier and more efficient.

Laboratory scale tests were carried out in a jig, in which the regular and irregular particles were enriched separately. The height of the working chamber of the device was 400 mm, the inner diameter of the rings: 100 mm, the height of the ring: 25 mm. The pulse frequency was 90 cycles per minute. Each product layer consisted of two rings: the two lowest rings constituted layer I (the bottom one), while the seventh and eighth rings were layer IV (the upper layer). Additionally, for comparison, enrichment tests were carried out for the material with natural content of regular and irregular particles. Experiments were carried out for various particle size fraction of material and various content of regular and irregular particles.

The second stage of experiments—semi-plant tests—were conducted for the feed prepared according to the patented technology for production of aggregates based on separation of narrow particle size fractions according to size and shape, and downstream densimetric enrichment of particles. The hard-enrichable particles separator (*SET*) is a device that utilizes water as a working medium for separation of the aggregate feed into two fractions differing from each other in terms of density and other physical and mechanical properties (Figure 2). The degree of aggregate separation is variable and depends on the amount of water used and an amount of individual size fractions of the aggregate. Primary operational characteristics of the separator parameters are presented in Table 2. Operational throughput of device was 2500 kg/h, and tests were conducted at high (serie I) and low (serie II) height of the separation threshold. In the serie I the position of the threshold in separator allowed for obtaining 75% of lower (heavier) and 25% of upper product. In the serie II of semi-plant investigations the share in separation products was reverse, i.e., 25% of lower product and 75% of upper one.

3.2. Research Programme and Scope of Analyses

The scheme of research programmes both in laboratory and in semi-plant scale is presented in Figure 3. The regular and irregular particles in each narrow size fraction were separated by classifying on a screen with slotted apertures, according to the standard PN-EN 933-3:2012. The prepared narrow size fractions were enriched in a jig and the idea of operation of such circuit was to eliminate equally settling particles prior to the gravitational separation process, which would have a negative effect on the separation sharpness in the jig. Detailed characteristics of the feed material for laboratory tests is presented in Table 3. Separation products were subjected to density, absorbability and porosity analyses. Density was determined with using of Archimedes law, while absorbability was established by weighing of wet and dry product and performing suitable calculations.

Separation products were subjected to density, absorbability and porosity analyses. Density was determined with using of Archimedes law, while absorbability was established by weighing of wet and dry product and performing suitable calculations. It is worth to mention here, that obtained absorbability values do not determine the real absorbability indices that can be determined according to relevant standards. In this case we rather mean the "processing" or "operational" absorbability, that is an amount of water that individual particle can absorb while it is in the jig during the process. The porosity, in turn, was determined manually, i.e., each particle was inspected and it was classified as a "porous particle" when pores and irregularities covered more than 50% of its surface. This porosity determined in this way can be rather understood as an external one, because this technique does not allow for determination of internal structure of individual particle.

Figure 2. Model of the SET separator (**left**), device in plant conditions (**right**).

Table 2. Characteristics of operational parameters of SET device.

Parameter	Unit	Value
Maximum throughput	[kg/h]	2750
Maximum water flow	[dm^3/h]	5500
Frequency of bellows pulsation	[1/s]	0.8–1.2
Jump of bellows	[mm]	50–140
Nominal power	[kW]	4
Dimensions of sieves	[mm]	150 × 2900

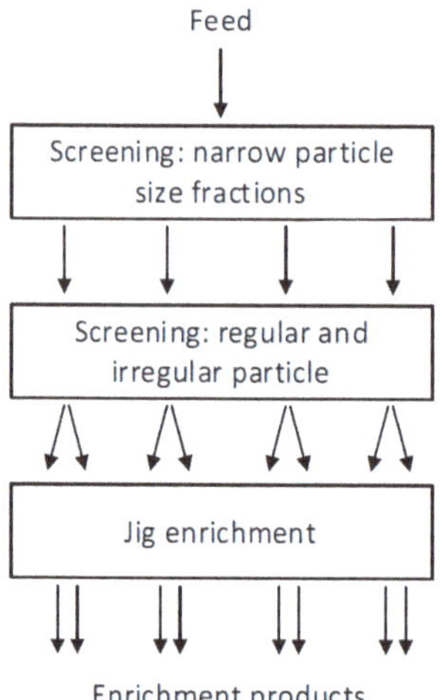

Figure 3. The screening and beneficiation scheme.

Table 3. Summary characteristics of individual tests.

Test Number	Type of Material	Particle Size, [mm]	Regular Particles Content in Feed, [%]	Irregular Particles Content in Feed, [%]
I	gravel	8–16	89	11
II	gravel	8–16	100	0
III	gravel	8–10	0	100
IV	gravel	8–10	100	0
V	gravel	6.3–8	0	100
VI	gravel	6.3–8	100	0

4. Results and Discussion

4.1. Laboratory Scale Tests

Each jig enrichment product in laboratory scale was assessed in terms of its water absorption and density and results are presented in Tables 4 and 5.

Table 4. Summary results of separation achieved in laboratory tests-density.

Test Number	Density, [g/cm^3]			
	Number of Layer in the Jig			
	I	II	III	IV
I	2.67	2.66	2.66	2.63
II	2.66	2.67	2.69	2.68
III	2.64	2.64	2.63	2.61
IV	2.74	2.66	2.65	2.62
V	2.64	2.62	2.60	2.60
VI	2.73	2.66	2.66	2.61

Table 5. Summary results of separation achieved in laboratory tests-absorbability.

Test Number	Absorbablity, [%]			
	Number of Layer in the Jig			
	I	II	III	IV
I	1.03	1.52	1.54	2.08
II	1.27	1.71	1.74	1.91
III	2.95	4.09	4.46	4.75
IV	2.55	3.34	3.54	3.96
V	2.97	3.21	3.47	4.05
VI	0.97	1.85	2.22	2.51

Analysis of results obtained for tests I and II (coarse aggregate in wide particle size fraction 8–16 mm) shows that absorbability achieved for the bottom layer I was the lowest. For consecutive higher layers it was increased, in total the difference between layers I and IV was over 1% for natural feed (containing 11% of irregular particles). For feed with regular particles, the respective difference was smaller, achieving approximately 0.8%. It is evident that the properties of individual layers are different when the feeds have 0 or 11% irregular particles, but the differences are relatively low, and don't exceed 0.5%. The values of densities in individual enrichment products are relatively small, however the upper layer (IV) demonstrates the highest difference between regular and irregular products of jig enrichment.

Analysis of specific enrichment products in narrow particle size fractions show that the achieved values of densities are much higher, both for coarse and fine particles, than in tests on a feed with a natural particle size composition. Differences in absorbability for two upper layer exceeds 1.5% which is approximately 5 to 6 times higher, comparing the

55

results obtained for wide particle size fraction 8–16 mm. Differences in absorbability for the two upper layer exceeds 1.5% which is approximately 5 to 6 times higher, compared to the results obtained for a wide particle size fraction 8–16 mm. Differences in densities are also higher in enrichment products treated in narrow particle size fractions. The highest difference, as high as 0.1%, was observed for the lowest layer I. The differences in densities for products from layers 2 to 4 from tests III and IV were as much as two times greater compared to analogous products from a feed with a particle size range of 8–16 mm.

Results of tests V and VI show that processing absorbability of regular particles was significantly lower than for irregular particles. In the case of density it can be observed that lower products contain the highest number of particles with highest density value. Together with increasing the number of separation product its density decreases. This phenomenon of density segregation is presented in Figure 4. Product from layer I with regular particles (left) had the highest content of quartz particles comparing to the other products, especially to irregular particles (test V), where a sandstone predominated.

Figure 4. Separation products: regular particles (**a**) (Test VI), irregular particles (**b**) (Test V).

Calculated averaged values of absorbability and density are presented in Table 6. Inspecting the table it can be seen that there is little difference between the values for irregular and regular products (tests I and II) when the material has a natural particle size composition. However, for material with a narrow particle size range (tests III and IV) the absorbability is much greater in material consisting of irregular particles than it is in material consisting of regular particles. This indicates that the application of the patented solution for improving the quality of beneficiated material—by splitting the feed into narrow size classes—is justified. No significant improvement in variation of density was observed. Summary of percentage contents of porous particles in each layer of enrichment were presented in Figure 5.

Table 6. Average values of absorbability and density for individual tests.

Test Number	Average Absorbability, [%]	Average Density, [g/cm³]
I	1.54	2.64
II	1.65	2.68
III	3.09	2.63
IV	1.38	2.67
V	3.38	2.61
VI	1.80	2.67

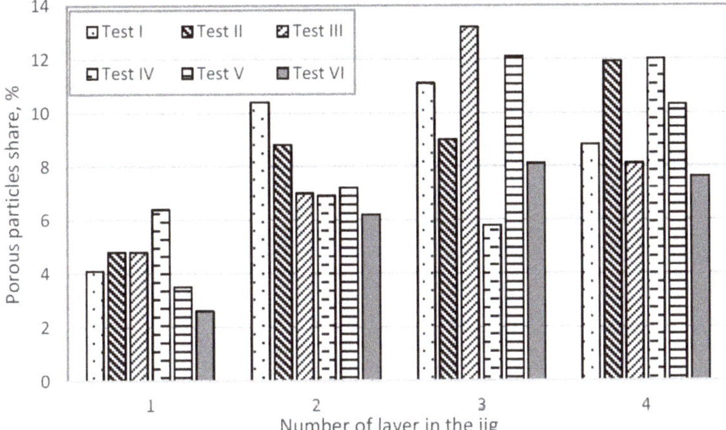

Figure 5. Characteristics of porosity of individual enrichment products for all laboratory tests.

The obtained results show that for tests I, III and V the highest percentage content of porous particles was observed in layer 2 and 3. It appears that this is valid for irregular particles products, regardless of the size range of the material. For tests V and VI, in turn, the content of irregular particle is the lowest for the bottom layer 1. It was caused through accumulation of the highest number of quartz particles with low porosity. It is also worth to mention that accumulation of porous particles in layer I was the lowest in all tests I–VI. It is also correlated with their density and absorbability. In layers 3 and 4, in turn, the highest content of porous particles can be observed.

Comparing the obtained results it can be concluded that the classification of the feed material into narrow particle size fractions and dividing them according to the shape brings better results in the jig enrichment process, due to the narrowing of the parameters of the aggregates and the elimination of equally settling particles.

Modeling results show that relationships between the number of separation product and its selected feature is in correlation, to some extent. Relationships between density/absorbability and number of the separation product (specific layer in the jig) do not show significant correlation for the entire set of data. There can be observed hyperbolic relationship, indeed, but statistical significance of this model is rather low (Figure 6).

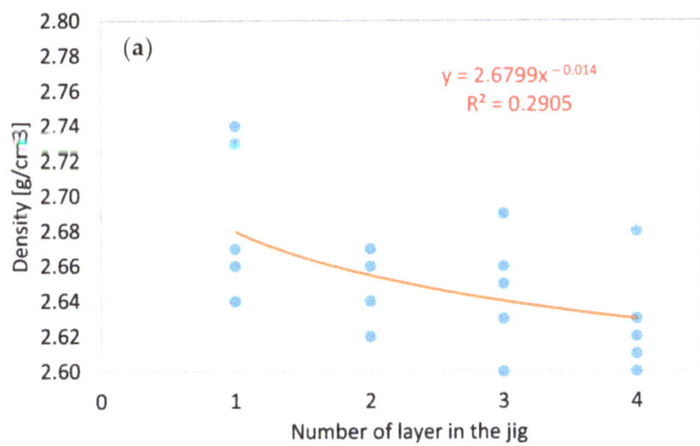

Figure 6. *Cont.*

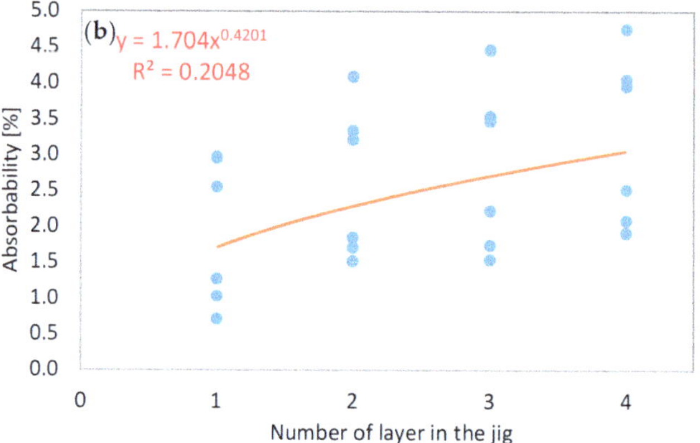

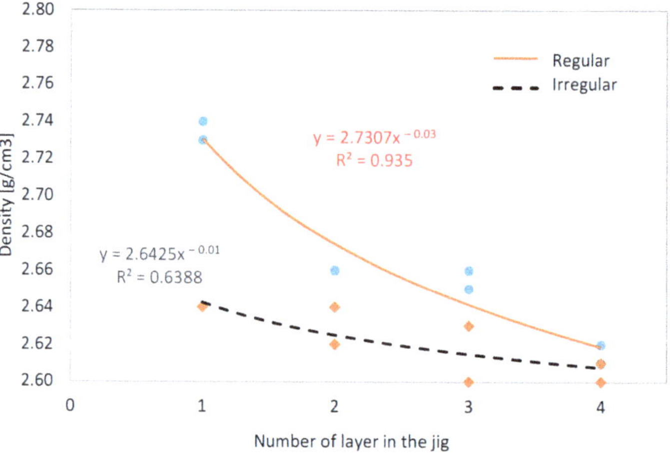

Figure 6. Model of separation for density (**a**) and absorbability (**b**).

More precise modeling results can be obtained when enrichment products obtained from regular and irregular feed material will be analyzed separately. Results for density was presented in Figure 7, while models for absorbability in Figure 8. Relationship between density and number of enrichment product for regular particles shows high level of accuracy ($R^2 = 0.935$) and is statistically significant. Lower statistical significance shows model for irregular particles but it is also significant on the probability level 95%. In the case of absorbability (Figure 6b) the model is statistically significant only for irregular particles. However both models show the similar type of relationship. Tendency in density decreasing together with increasing the number of layer describes well a hyperbola, while absorbability can be characterized through exponential model, with the power at x less than one.

Figure 7. Model of separation density calculated separately for regular and irregular particles.

Density models show that for regular particles the power at x is greater in absolute numbers, but the second coefficient shows no impact on the shape of particles. Models of absorbability show that power at x is also greater for regular particles. In this case this value is greater than zero. Higher impact of the second coefficient can be observed, and it is higher for irregular particles (2.9632) than for regular ones (1.4325). In general, the modeling results justify the approach consisting in run of jig beneficiation process separately for regular and irregular particles. Achieved accuracies of fitting to operational data were higher both for density and absorbability. It could be then easier to predict the outcomes for such processes, not to mention the main achievement–improvement of qualitative characteristics for products enriched in such a manner.

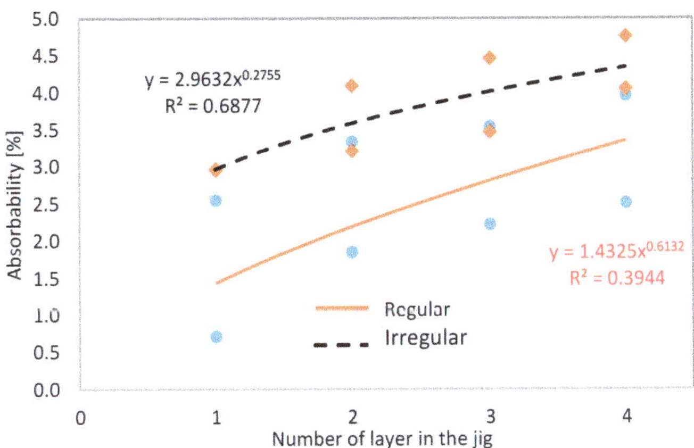

Figure 8. Model of separation density calculated separately for regular and irregular particles.

4.2. Semi-Plant Scale Tests

Results of tests conducted in semi-plant scale are presented in Tables 7 and 8. The test series with a high threshold level shows that differentiation of products both in terms of absorbability and density is not very significant. It is probably due to relatively wide range of particle size (10–16 mm) of the feed material.

It has turned out that preparation of the aggregate for narrower particle size fractions by screening and size separating gave more favorable enrichment results. It was confirmed in results of tests III-VI, especially for aggregates with irregular particles (tests III and V). Enrichment products, depending the number of a layer, were also differentiated in terms of density.

Table 7. Summary characteristics of individual tests in separator SET (high threshold).

Test Number	Layer of Product	Absorbability, [%]	Density, [g/cm^3]
I	lower	1.32	2.66
	upper	1.71	2.63
II	lower	1.49	2.69
	upper	1.83	2.68
III	lower	3.58	2.64
	upper	4.51	2.61
IV	lower	3.18	2.68
	upper	3.81	2.62
V	lower	2.26	2.63
	upper	3.67	2.61
VI	lower	1.26	2.70
	upper	2.20	2.61

Results of experiments conducted at low threshold also showed no significant differences in terms of density. There were observable significant variations in enrichment of narrow particle size fractions, especially fo regular particles (test IV and VI), in turn. These results confirm the purposefulness of suitable preparation of the feed material for enrichment process, it is expected to obtain an appropriate differentiation of products in terms of water absorption and density. On the example of test I (Table 7) it can be noticed that about 65% of flat particles were accumulated on the surface (Figure 9), while the remaining of irregular particles were allocated deeper in both layers. The height of the upper layer was 20 mm.

Table 8. Summary characteristics of individual tests in separator SET (low threshold).

Test Number	Layer of Product	Absorbability, [%]	Density, [g/cm^3]
I	lower	1.11	2.67
	upper	1.93	2.65
II	lower	1.29	2.66
	upper	1.91	2.68
III	lower	3.05	2.65
	upper	4.66	2.62
IV	lower	2.49	2.75
	upper	3.88	2.66
V	lower	2.89	2.66
	upper	3.49	2.60
VI	lower	0.68	2.73
	upper	2.12	2.65

Figure 9. Upper layer in the *SET* separator for the test with irregular particles.

Shape investigations of separation products were carried out by means of 3D Keyence VHX-7000 microscope. The figures map the profiles of the projection surface (base of particle) in space after transformation of an image into 3D by the computer (Figure 10). A particle is considered irregular if a profile of the length of its base is three times larger than the profile of its width or height. The analysis proves that more flat particles accumulate in the top layer.

Further investigations included comparative analysis of enrichment products obtained in the serie II (low threshold) with typical aggregates before enrichment process in SET device. Grinding resistance (Los Angeles) according to the standard PN-EN 1097-2, as well as micro-Deval (PN-EN 1097-1), were determined. Results are presented in Table 9.

Analysis of results presented in Table 9 indicates that the raw aggregate with 11% of irregular particles is characterized by the least favorable parameters (LA = 40, MDE = 30). The most favorable parameters, in turn, were obtained for enrichment product of the SET device, achieving LA = 30 and MDE = 10. Application of enrichment process makes it possible to increase significantly quality of physical and mechanical parameters of products, and at the same time increases potential commercial and industrial utilization of such products. It is especially significant in exploitation of poor or lower quality deposits or utilization of aggregates from recycling or tailings.

Figure 10. Exemplary analysis of irregular (**left**) and regular (**right**) particle in 3D Keyence VHX-7000 microscope.

Table 9. Results of Los Angeles and micro-Deval indices for a gravel aggregate in particle size fraction 10–14 mm at different stages of enrichment.

Gravel Aggregate 10–14 mm	Los Angeles Index, LA [%]	Micro-Devala Index, M_{DE} [%]
Raw material (typical) with 11% of irregular particles content	36.7 category LA40	29.8 category M_{DE} 30
Raw material without irregular particles	31.9 category LA35	17.6 category M_{DE} 20
Product enriched in SET device, without regular particles (low threshold)	29.5 category LA30	9.8 category M_{DE} 10

5. Conclusions

The aim of the paper was to demonstrate how the efficiency of a jig separation process for aggregate minerals could be improved. Thanks to the application of a patented classification circuit prior to jigging, the absorbability and density of individual products in specific layers in the jig were diverse on average for 0.8 and 0.5% respectively, comparing to values achieved for material without prior pretreatment.

In terms of water absorption, very favorable results (significant diversification of results) were obtained especially for narrow particle size fractions of separation products, where the difference between individual layers was greater than 2% (laboratory tests V and VI). Similar differences were observed for the density. The particles with the highest density have accumulated in the lower layers, which was due to the fact that in this layer the majority of quartz particles of low porosity was also present. It is also worth noting that in all tests, the least porous particles accumulated in layer 1, which also correlates with the density of these grains and their water absorption.

It should be noted, however, that greater differentiation between the enrichment products was obtained for using of a laboratory ring jig in comparison with the semi-plant *SET* device. This difference mainly results from the number of layers and the longer duration of enrichment process (5 min) in laboratory jig. In the *SET* device, in turn, the material has been enriching for several dozen seconds. It should be also taken into account that the *SET* device is a prototype that operates together with a set of screens separating the aggregates into narrow particle fractions according to size and shape, constituting the patented invention number PL 233318B1.

Considering the obtained results, it can be concluded that screening of feed material into narrow particle size fractions and the separation of these fractions also in terms of shape brings more favorable results of jig aggregate enrichment, due to the narrowing of the parameters of the aggregates and the elimination of equally settling particles.

The obtained results of the investigations show that jig beneficiation of aggregates in the proposed circuit may result in more efficient process performance, and also leads to more efficient use of raw materials and less waste production. It needs to be pointed out that the obtained differences in density and absorbability are not very large. However, a suitable design of the circuit and process operation may be effective in producing a more homogenous aggregate product in terms of their physical and mechanical properties. An application of jig beneficiation in the mineral aggregates production industry may help both in efficient separation of the final products and in elimination of fractions of particles with an abnormal density and/or absorbability from the final products, as well as decreasing the content of irregular particles. An overarching aim of the approach is increasing the strength properties of aggregate products, which attracts a higher value in many sectors of the building industry. A good example can be results of comminution resistance (Los Angeles index) and micro-Deval abrasion resistance. The raw aggregate with natural content of irregular particles (app. 11%) was accounted to lower categories according to LA and MDE. After enrichment process the product was characterized by higher categories, at the same time gaining the higher quality.

Patents: Two patents granted in Poland were utilized in the paper:

Author: Gawenda T. Title: Układ urządzeń do produkcji kruszyw foremnych, AGH w Krakowie.

Patent No. PL233689 granted on 8.07.2019.

Authors: Gawenda T., Saramak D., Naziemiec Z. Title: *Układ urządzeń do produkcji kruszyw oraz sposób produkcji kruszyw.* AGH w Krakowie, Patent No. PL 233318B1 granted on 7. 06. 2019.

Norms and standards: PN-EN 933-3:2012. *Badania geometrycznych właściwości kruszyw-część 3: Oznaczanie kształtu ziaren za pomocą wskaźnika płaskości.*

Author Contributions: Conceptualization, T.G. and Z.N.; methodology, T.G., D.S.; formal analysis, T.G., D.S., A.S., Z.N.; investigation, T.G., D.S., A.S., Z.N.; writing—original draft preparation, T.G.; writing—review and editing, D.S., A.S.; visualization, D.S., A.S.; supervision, T.G. All authors have read and agreed to the published version of the manuscript.

Funding: The paper is the effect of completing the NCBiR Project, contest no. 1 within the subaction 4.1.4 "Application projects" POIR in 2017, entitled "Elaboration and construction of the set of prototype technological devices to construct an innovative technological system for aggregate beneficiation along with tests conducted in conditions similar to real ones". The Project is co-financed by the European Union from sources of the European Fund of Regional Development within the Action 4.1 of the Operation Program Intelligent Development 2014–2020.

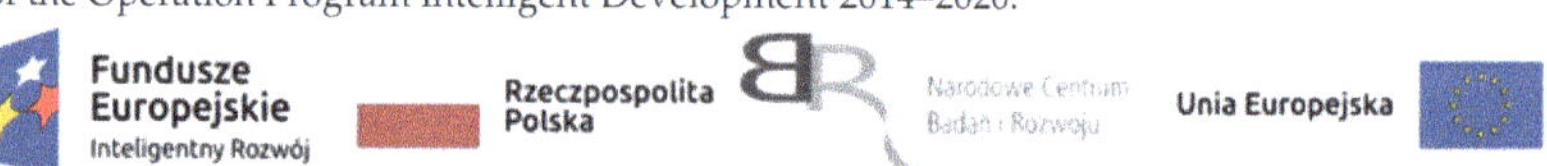

Conflicts of Interest: The authors declare no conflict of interest.

References

1. Heyduk, A.; Pielot, J. Economical Efficiency Assessment of an Application of On-line Feed Particle Size Analysis to the Coal Cleaning System in Jigs. *Inżynieria Miner. J. Pol. Miner. Eng. Soc.* **2014**, *2*, 217–228.
2. Saramak, A.; Naziemiec, Z. Determination of dust emission level for various crushing devices. *Min. Sci.* **2019**, *26*, 45–54. [CrossRef]
3. Saramak, A.; Naziemiec, Z.; Saramak, D. Analysis of noise emission for selected crushing devices. *Min. Sci.* **2016**, *23*, 145–154. [CrossRef]
4. Ambróst, W. Jigging: A review of fundamentals and future directions. *Minerals* **2020**, *10*, 998. [CrossRef]
5. Falconer, A. Gravity separation: Old technique/new methods. *Phys. Sep. Sci. Eng.* **2003**, *12*, 31–48. [CrossRef]

6.   Biswajit Sarkar, B.; Sekhar, S.C.; Das, A. *Advanced Gravity Separation*; Singh, R., Das, A., Goswani, N.G., Eds.; NML: Jamshedpur, India, 2007; p. 831007.
7.   Boron, S.; Pielot, J.; Wojaczek, A. Coal cleaning in jig systems—Profitability assessment. *Miner. Resour. Manag.* **2014**, *30*, 67–82.
8.   Cierpisz, S. A dynamic model of coal products discharge in a jig. *Miner. Eng.* **2017**, *105*, 1–6. [CrossRef]
9.   Głowiak, S. Wpływ składu ziarnowego nadawy na skuteczność wzbogacania w osadzarce. In Proceedings of the XV APPK, Szczyrk, Poland, 2–4 June 2009; pp. 37–50.
10. Gawenda, T. *Zasady Doboru Kruszarek Oraz Układów Technologicznych w Produkcji Kruszyw Łamanych*; Monography no. 304; AGH Publishing House: Cracow, Poland, 2015.
11. Neumann, T.; Snoby, R.J.; Strangalies, W. The fractionized separation of impurities out of sand and small gravel with alljig-fine grain jigs. *Aufbereit. Technik.* **1995**, *36*, 562–567.
12. Mesters, K.; Kurkowski, H. Density separation of recycling building materials by means of jig technology. *Aufbereit. Technik.* **1997**, *38*, 536–542.
13. Phengsaart, T.; Ito, M.; Hamaya, N.; Tabelin, C.B.; Hiroyoshi, N. Improvement of jig efficiency by shape separation, and a novel method to estimate the separation efficiency of metal wires in crushed electronic wastes using bending behavior and entanglement factor. *Miner. Eng.* **2018**, *129*, 54–62. [CrossRef]
14. Ito, M.; Saito, A.; Murase, N.; Phengsaart, T.; Kimura, S.; Tabelin, C.B.; Hiroyoshi, N. Development of suitable product recovery systems of continuous hybrid jig for plastic-plastic separation. *Miner. Eng.* **2019**, *141*, 105839. [CrossRef]
15. Cazacliu, B.; Sampaio, C.H.; Miltzarek, G.; Petter, C.; Le Guen, L.; Paranhos, R.; Huchet, F.; Kirchheim, A.P. The potential to using air jigging to sort recycled aggregates. *J. Clean. Prod.* **2014**, *66*, 46–53. [CrossRef]
16. Sampaio, C.H.; Ambrós, W.M.; Miranda, L.R.; Gerson, L.; Miltzarek, G.M.; Kronbauer, M.A. Improve the quality of recycled aggregate concrete by sorting in air jig. In Proceedings of the III Progress of Recycling in the Built Environment, São Paulo, Brazil, 3–5 August 2015.
17. Stempkowska, A.; Gawenda, T.; Naziemiec, Z.; Ostrowski, K.; Saramak, D.; Surowiak, A. Impact of the geometrical parameters of dolomite coarse aggregate on the thermal and mechanic properties of preplaced aggregate concrete. *Materials* **2020**, *13*, 4358. [CrossRef] [PubMed]
18. Surowiak, A.; Gawenda, T.; Stempkowska, A.; Niedoba, T.; Nad, A. The Influence of Selected Properties of Particles in the Jigging Process of Aggregates on an Example of Chalcedonite. *Minerals* **2020**, *10*, 600. [CrossRef]
19. Hori, K.; Tsunekawa, M.; Hiroyoshi, N.; Ito, M. Optimum water pulsation of jig separation for crushed plastic particles. *Int. J. Miner. Process.* **2009**, *92*, 103–108. [CrossRef]
20. Dos Santos, I.L.; Frantz, L.V.; Masuero, A.B. Influence of hydraulic jigging of construction and demolition waste recycled aggregate on hardened concrete properties. *Rev. IBRACON Estruturas Mater.* **2021**, *14*, 14314. [CrossRef]
21. Burt, R.O. *Gravity Concentration Technology*; Elsevier: Amsterdam, The Netherlands, 1984.
22. Ottley, D.J. Gravity Concentration In Modern Mineral Processing. In *Mineral Processing at a Crossroads*; Wills, B.A., Barley, R.W., Eds.; Springer: Berlin/Heidelberg, Germany, 1986; Volume 117. [CrossRef]
23. Naziemiec, Z.; Gawenda, T. Badanie procesu kruszenia z zamkniętym obiegiem. In Proceedings of the Kruszywa Mineralne 2007 Surowce—Rynek—Technologie—Jakość, Szklarska Poręba, Poland; 2007; pp. 107–116.
24. Wills, B.A. *Mineral Processing Technology*, 6th ed.; Pergamo Press: Oxford, UK, 2006.
25. Brożek, M.; Surowiak, A. Argument of separation at upgrading in the JIG. *Arch. Min. Sci. Arch. Górnictwa* **2010**, *55*, 21–40.
26. Gawenda, T.; Saramak, D.; Nad, A.; Surowiak, A.; Krawczykowska, A.; Foszcz, D. Badania procesu uszlachetniania kruszyw w innowacyjnym układzie technologicznym. In Proceedings of the XIX Conference Kruszywa Mineralne Surowce—Rynek—Technologie—Jakość, Kudowa-Zdrój, Poland, 25–28 September 2019; pp. 65–76.

Article

Multidimensional Optimization of the Copper Flotation in a Jameson Cell by Means of Taxonomic Methods

Tomasz Niedoba [1,*], Paulina Pięta [2], Agnieszka Surowiak [1] and Oktay Şahbaz [3]

[1] Department of Environmental Engineering, Faculty of Mining and Geoengineering, AGH University of Science and Technology, al. Mickiewicza 30, 30-059 Kraków, Poland; asur@agh.edu.pl
[2] JSW Innowacje S.A., ul. Paderewskiego 41, 40-282 Katowice, Poland; ppieta@jswinnowacje.pl
[3] Faculty of Engineering, Kütahya Dumlupinar University, Evliya Çelebi Yerleşkesi Tavşanlı Yolu 10.km, 43100 Kütahya, Turkey; oktay.sahbaz@dpu.edu.tr
* Correspondence: tniedoba@agh.edu.pl; Tel.: +48-126172056

Abstract: Three factors were measured in the flotation process of copper ore: the copper grade in a concentrate (β), the copper grade in tailings (ϑ), and the recovery of copper in a concentrate (ε). The experiment was conducted by means of a Jameson cell. The factors influencing the quality of the process were the particle size (d), the flotation time (t), the type of collector (k), and the dosage of the collector (s). The considered vector function is then ($\beta(d, t, k, s)$, $\vartheta(d, t, k, s)$, $\varepsilon(d, t, k, s)$). In this work, the optimization was based on determining the values of the adjustable factors (d, t, k, s). The goal was to obtain the possibly highest values of the functions β and ε (maximum) with the possibly lowest values of the function ϑ (minimum). To this end, taxonomic methods were applied. Thanks to the applied method, the optimum—with the adopted assumptions—was found. The presented methodology can be successfully applied in the search for the optima in a variety of technological processes.

Keywords: flotation; copper ore; lithology; flotation agents; particle size distribution; taxonomic methods

Citation: Niedoba, T.; Pięta, P.; Surowiak, A.; Şahbaz, O. Multidimensional Optimization of the Copper Flotation in a Jameson Cell by Means of Taxonomic Methods. *Minerals* **2021**, *11*, 385. https://doi.org/10.3390/min11040385

Academic Editor: Dave Deglon

Received: 16 February 2021
Accepted: 31 March 2021
Published: 3 April 2021

Publisher's Note: MDPI stays neutral with regard to jurisdictional claims in published maps and institutional affiliations.

1. Introduction

The main operation of copper ore beneficiation, after its preparation in the processes of fragmentation and classification, consists in the application of the flotation process in the multi-stage final grinding and cleaning systems. Polish copper ore is characterized by three main lithological fractions which require a different way of beneficiation, with flotation as the second stage of the process. The main lithological fractions are presented in Table 1 showing the characteristics of the feed entering the technological system. The percentage shares of all lithological types vary depending on the region of occurrence. The content of copper in the ore used as feed for the process of beneficiation in processing plants changes depending on the lithological content of the feed, which is closely related to its region of occurrence. Therefore, the technology of copper ore beneficiation depends on its lithological composition. For this reason, the general ore processing variant cannot be used as its mineralogical and qualitative composition changes in the same way as mining and geological conditions of ore occurrence change. Apart from copper, the feed for beneficiation contains associated elements, i.e., silver, gold, platinum, and others, which also occur in varying amounts and are associated with the lithological type. The occurrence of three lithological types of Polish copper ore depostis significantly hinders the process of output beneficiation due to the diversity of their mineralogical and physico-chemical properties. The decrease in the size of ore-bearing particles observed in recent years makes it necessary to perform the grinding in finer size particle distributions with the aim to release copper-bearing particles. However, flotation of very fine particles is difficult to perform in efficient way [1]. Therefore, it is necessary to use a new generation of machines with adequately selected bubble size distribution, which enable the adhesion of extremely fine particles [2]. In general, the processing of ore and the production of a concentrate for metallurgical processes of suitable quality requires that the process of beneficiation is conducted with utmost care in order to ensure optimal quantitative and qualitative parameters of the produced feed for metallurgy [1]. From the perspective of the assessment of the processing plant's final product, the most important assessment indicators comprise the content of copper in the concentrate, the waste, as well as the yield of copper in the concentrate. The process of mineral flotation depends on many factors, i.e., the minerals' nature and structure (mineralogy, morphology, and particle size), water chemistry, bubble size and velocity, flotation time, hydrodynamic properties, pulp potential and pH, pulp density, air flow rate, as well as reagent types and dosages [1–5]. To achieve the best possible indicators, the process of flotation is conducted with the optimization of some technical and technological parameters.

Table 1. Mineralogical composition of lithological types of Polish copper ores [6].

Lithological Type of Copper Ore	Content of Selected Metals in Lithological Types		Prevalent Copper-Bearing Minerals
carbonates	Cu (%)	1.69	chalcocite in combination with digenite, bornite, covellite and chalcopyrite
	Ag (g/t)	54	
shales	Cu (%)	6.02	chalcocite-bornite and bornite-chalcopyrite minerals
	Ag (g/t)	188	
sandstones	Cu (%)	1.29	bornite-chalcopyrite and chalcocite-bornite minerals
	Ag (g/t)	30	

Many studies on ore flotation are available in the literature. Most of them deal with various optimization issues. With regard to copper ores, many papers discuss the problem of selecting appropriate reagents and their dosages. The use and selection of new kinds of reagents for the process was the topic of the studies presented in [7–11]. The introduction of seawater was presented in [12]. The effect of desliming on flotation efficiency was investigated by [13]. Podariu et al. discussed the role of metallic electrodes in the process [14]. The problem of bubble size distribution as well air rate and froth depth were the object of interest in [2,15]. The application of ultrasound at various stages of the copper flotation process was discussed in [16]. One of the main factors for evaluating the quality of the process is the selectivity index. A study on the impact of the process parameter modification was presented in [17,18]. The surface oxidation level was investigated in [19]. Furthermore, various attempts in the modeling of the whole process or parts of it, introducing different types of algorithms, were presented in many papers [20–26]. We have also conducted many studies on copper ore processing and its optimization. Many different methods were applied for this purpose. A parametric optimization in mixed copper ores flotation was presented in [27]. A geometrical approach was the subject presented in [28]. A combined approach consisting of neural networks and evolutionary algorithms was shown in [29]. Non-classical statistical methods, such as kernel methods, Fourier series method, or non-parametric statistical methods were introduced in [30]. Applications of ANOVA (Analysis of Variance) in mineral processing, including also copper flotation were discussed in [31]. The initial studies of the copper flotation process conducted in a Jameson cell was the subject presented in [32]. To this end, we used taxonomic methods, which are an innovative approach to optimize the process. Copper grade in concentrate (β), copper grade in tailings (ϑ), and copper recovery in a concentrate (ε) were selected as factors for the evaluation of the flotation performance (performance indicators). In this study, adjustable factors that influence flotation quality are the particle size (d), the separation time (t), the collector type (k), and the collector dosage (s).

2. Experiment

2.1. Laboratory Investigation

The experimental research was conducted through a Jameson cell. It is a pneumatic flotation device in which pressurized, naturally aspirated air is dispersed. It is responsible for the mixing of the suspension. The device consists of two main parts, which are the downcomer and the separation tank. Conditioned particles are pumped to the nozzle at the top of the downcomer to create a high-pressure water jet, and the air is sucked into the downcomer. This water jet is responsible for producing a high-intensity mixing and fine bubbles. Thus, the downcomer becomes the first contact point of particles and air bubbles. Micro-events of flotation occur in the downcomer, and hydrophobic particles become attached to air bubbles. A bubbly mixture is discharged to the separation tank from the downcomer. The separation tank provides a suitable environment for the separation of hydrophilic particles from the particle-laden bubbles. Hydrophobic particles–bubbles aggregates are raised to the froth zone. There is a water washing system, which positively impacts the selectivity of the process [33–38].

During this operation, fine bubbles increase the collision between bubbles and particles and improve the flotation kinetics. This characteristic lowers the requirements regarding particle retention time and makes it possible to decrease the Jameson cell height compared to traditional flotation columns [39–42].

The investigated material was Polish carbonate copper ore. The initial copper grade in the feed equaled 1.5%. From the lithological point of view, it contained minerals, such as carbonates (dolomite, calcite)—about 72%, shale minerals—about 16%, sulfates (gypsum, anhydrite)—5%, quartz—3%, copper sulfides—3% and organic substance—0.5%. The Jameson cell scheme is presented in Figure 1.

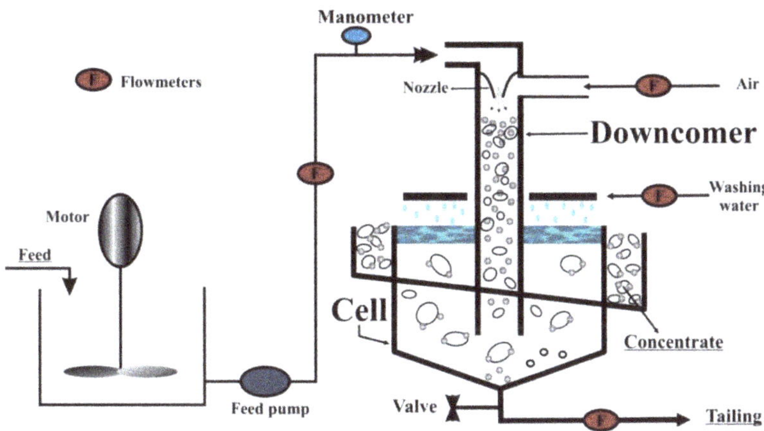

Figure 1. Scheme of a Jameson cell [39]

The parameters of the flotation machine were the following:

- separation tank diameter and height 200 mm and 900 mm, respectively;
- downcomer diameter and length: 0.020 m and 1.8 m, respectively;
- nozzle diameter 0.005 m;
- conditioning tank volume: 0.1 m^3;
- downcomer plunging length that is the depth to which the end of the downcomer is immersed in the separation tank: 0.5 m;
- feed rate and air rate: 100 cm^3/s.

The investigation was based on the changes in the course of the process, caused by the changes in the individual factors (adjustable variables). Particle fractions −20,

20–40, and 40–71 μm were prepared for the tests. For each level of the experiment, it was necessary to repeat the process in order to verify the adequacy of the results. The results of the laboratory experiments were significant and the values of errors did not exceed the acceptable limits (<5%) which were evaluated using the standard deviation. It was assumed that the maximum time of flotation would amount to 30 min. The concentrate was collected selectively after 1, 2, 4, 6, 9, 12, 17, 22, and 30 min. As a result, it was possible to analyze the kinetics of the separation as well as the influence of time on the effects of beneficiation. The solids grade in the Jameson cell was maintained at a constant level of 2%. The Nasfroth frother was added to the amount of 50 g/t. The final stage was to determine the copper content in the separation products with the use of the XRF methodology, which made it possible to calculate process factors, such as the copper grade in the concentrate β, the copper grade in the tailings ϑ, and the copper recovery in the concentrate ε.

The variables were selected on the basis of previous experiments which showed that these factors are strongly related to the efficiency of the flotation process [34,35]. The values of these adjustable factors are outlined in Table 2.

Table 2. Characteristics of adjustable factors.

Particle Size Fraction d (μm)	Collector Type k	Collector Dosage s (g/t)	Time t (min)
0–20 20–40 40–71	Aqueous solution of ethyl sodium xanthate—E Aqueous solution of isobutyl sodium xanthate—I	100 150	1, 2, 4 6, 9, 12 17, 22, 30

For each determined value of the adjustable parameters (d, t, k, s) five measurements of researched flotation factors were performed, which results in a vector (β, ϑ, ε).

The averaged results of measurements and calculations are presented in Tables 3–5.

Table 3. Results of measurements for particle size fraction 0–20 (μm).

Time t (min)	E						I					
	100 (g/t)			150 (g/t)			100 (g/t)			150 (g/t)		
	β	ϑ	ε	β	ϑ	ε	β	ϑ	ε	β	ϑ	ε
1	0.145	0.024	0.068	0.113	0.018	0.152	0.166	0.026	0.152	0.171	0.023	0.122
2	0.171	0.017	0.356	0.121	0.013	0.402	0.184	0.019	0.416	0.157	0.018	0.359
4	0.160	0.012	0.572	0.113	0.010	0.594	0.175	0.013	0.613	0.150	0.014	0.562
6	0.141	0.008	0.709	0.102	0.007	0.706	0.156	0.009	0.737	0.137	0.009	0.706
9	0.126	0.006	0.782	0.092	0.006	0.772	0.142	0.007	0.811	0.124	0.007	0.785
12	0.115	0.006	0.802	0.086	0.005	0.798	0.131	0.006	0.833	0.115	0.006	0.809
17	0.108	0.005	0.832	0.081	0.004	0.830	0.123	0.006	0.856	0.107	0.005	0.835
22	0.104	0.005	0.858	0.077	0.004	0.839	0.119	0.005	0.882	0.103	0.005	0.856
30	0.097	0.004	0.876	0.072	0.004	0.856	0.112	0.004	0.895	0.097	0.004	0.870

Table 4. Results of measurements for particle size fraction 20–40 (μm).

Time t (min)	E						I					
	100 (g/t)			150 (g/t)			100 (g/t)			150 (g/t)		
	β	ϑ	ε	β	ϑ	ε	β	ϑ	ε	β	ϑ	ε
1	0.076	0.024	0.065	0.065	0.026	0.017	0.063	0.018	0.049	0.167	0.021	0.137
2	0.076	0.021	0.188	0.081	0.023	0.148	0.075	0.016	0.220	0.135	0.013	0.336
4	0.077	0.019	0.301	0.078	0.021	0.256	0.082	0.013	0.403	0.121	0.013	0.505
6	0.075	0.017	0.414	0.079	0.019	0.372	0.074	0.011	0.509	0.114	0.010	0.633
9	0.076	0.015	0.506	0.077	0.017	0.453	0.068	0.010	0.569	0.105	0.009	0.706
12	0.073	0.014	0.545	0.076	0.016	0.501	0.063	0.010	0.589	0.097	0.008	0.733
17	0.073	0.012	0.602	0.073	0.015	0.538	0.059	0.009	0.615	0.091	0.008	0.754
22	0.074	0.011	0.645	0.073	0.014	0.577	0.058	0.009	0.639	0.088	0.007	0.777
30	0.071	0.011	0.677	0.071	0.013	0.610	0.055	0.009	0.658	0.083	0.007	0.796

Table 5. Results of measurements for particle size fraction 40–71 μm.

Time t (min)	E						I					
	100 (g/t)			150 (g/t)			100 (g/t)			150 (g/t)		
	β	ϑ	ε	β	ϑ	ε	β	ϑ	ε	β	ϑ	ε
1	0.075	0.019	0.032	0.064	0.020	0.065	0.092	0.024	0.024	0.075	0.025	0.041
2	0.057	0.017	0.142	0.065	0.018	0.193	0.068	0.022	0.132	0.068	0.023	0.144
4	0.057	0.016	0.253	0.068	0.016	0.326	0.068	0.020	0.238	0.063	0.022	0.226
6	0.057	0.014	0.372	0.072	0.013	0.467	0.065	0.019	0.328	0.060	0.021	0.299
9	0.058	0.013	0.462	0.069	0.011	0.548	0.062	0.018	0.386	0.056	0.020	0.348
12	0.058	0.012	0.515	0.067	0.010	0.592	0.059	0.017	0.413	0.055	0.020	0.382
17	0.056	0.011	0.546	0.065	0.010	0.630	0.057	0.017	0.446	0.055	0.019	0.419
22	0.055	0.011	0.580	0.064	0.009	0.667	0.056	0.016	0.468	0.055	0.018	0.443
30	0.054	0.010	0.613	0.062	0.008	0.700	0.053	0.016	0.488	0.052	0.018	0.463

Table 3 presents the results experimentally obtained for the particle size fraction −20 μm for both reagent types (E, I) at doses of 100 and 150 (g/t), depending on the flotation time.

Table 4 shows analogous results, but for the fraction 20-40 μm. Similarly, as in the case of the finest size fraction, it is also difficult to determine the optimal point of the process in this case, taking into consideration the values of all three technological indicators.

Table 5 shows the results obtained for the fraction 40–71 μm. The conclusions are similar. Multivariate statistical methods must be used in order to determine the optimal conditions. This paper proposes the application of taxonomic methods, whose use is innovative in the context of problems related to the the processing of raw materials.

2.2. Methodology of Taxonomic Methods

2.2.1. Theoretical Background

The selected taxonomical methods found wide application in various scientific disciplines [43–46], because their major advantages are universality, simplicity of calculations, and simple interpretation of the results. The taxonomic factors allow to replace the description of the considered multi-feature object by means of one synthetic variable. The complex structure of the flotation process as well as the changeability of the investigated copper ore make it necessary to apply multidimensional methods for data analysis [47–50]. The basis to conduct the multidimensional comparison analysis is a matrix of diagnostic features X (1), which is then standardized and transformed into a synthetic factor Z (2). All considered situations are put in order in a linear way with consideration of the positive influence (stimulants) and the negative influence (destimulants) on the researched phenomenon. Then the surrogate variable is introduced as the distance between the objects which allow to evaluate the phenomenon. The development of the taxonomy caused the introduction of various factors and methods of variable normalization [51].

$$X = \begin{bmatrix} x_{11} & x_{12} & \cdots & x_{1l} \\ x_{21} & x_{22} & \cdots & x_{2l} \\ \cdots & \cdots & \cdots & \cdots \\ \cdots & \cdots & \cdots & \cdots \\ x_{n1} & x_{n2} & \cdots & x_{nl} \end{bmatrix} \tag{1}$$

$$Z = \begin{bmatrix} z_{11} & z_{12} & \cdots & z_{1l} \\ z_{21} & z_{22} & \cdots & z_{2l} \\ \cdots & \cdots & \cdots & \cdots \\ \cdots & \cdots & \cdots & \cdots \\ z_{n1} & z_{n2} & \cdots & z_{nl} \end{bmatrix} \tag{2}$$

Among the taxonomical methods many factors can be used. In this work, the Euclidean distance e_i was used, whose general formula is presented by Equation (3).

$$e_i = \sqrt{\sum_{j=1}^{l} (1 - z_{ij})^2} \text{ for } i = 1, \ldots, n.$$

(3)

where:

i—number of the row;
j—number of the column;
n—number of investigated variables (flotation tests);
l—number of variables (process evaluation factors);

$$z_{ij} = \frac{x_{ij}}{x_{jmax}}, \ x_{jmax} = \max_i (x_{ij}) \text{ standardized value.}$$

(4)

For such determined values of z_{ij} the values $e_1, e_2, \ldots, e_n$ were calculated by means of Equation (3). The smallest value allowed us to determine the optimal values of the considered factors.

The precise description of how to conduct the investigation by means of taxonomic methods can be found in [45,47,51].

2.2.2. Application

The multidimensional projection considered in this work takes the following form:

$$f : (d,t,k,s)(\beta(d,t,k,s), \vartheta(d,t,k,s), \varepsilon(d,t,k,s))$$

(5)

where values of variables (d, t, k, s) are accepted in accordance with the values proposed in Table 2.

Next, the optimization of the flotation process is performed. It is based on the determination of such values of adjustable factors (d, t, k, s) for which the functions β and ε assume simultaneously the biggest values and the function ϑ the smallest one.

Because of the fact that it is required that the variables β and ε reach the highest possible values in order to be qualified as flotation process stimulants, while the variable ϑ is treated as a destimulant. According to the taxonomic methods, destimulants should be transferred to become stimulants. That is why a new variable, $\frac{1}{\vartheta_i}$, is introduced instead of the variable ϑ.

In order to enable the comparison of various values, they need to be normalized first. It can be done by the introduction of new variables, according to Equations (6)–(8).

$$\tilde{\varepsilon}_i = \frac{\varepsilon_i}{\max_j \varepsilon_j}$$

(6)

$$\tilde{\beta}_i = \frac{\beta_i}{\max_j \beta_j}$$

(7)

$$\tilde{\vartheta}_i = \frac{\frac{1}{\vartheta_i}}{\max_j \frac{1}{\vartheta_j}}$$

(8)

Selection of the optimal adjustable variables is performed using the function of minimization

$$F(\beta(d,t,k,s), \vartheta(d,t,k,s), \varepsilon(d,t,k,s))$$

(9)

where

$$F(\beta_i, \vartheta_i, \varepsilon_i) = \sqrt{(1 - \tilde{\varepsilon}_i)^2 + \left(1 - \tilde{\beta}_i\right)^2 + \left(1 - \tilde{\vartheta}_i\right)^2}$$

(10)

where $\tilde{\varepsilon}_i$, $\tilde{\beta}_i$, $\tilde{\vartheta}_i$ are provided by Equations (6)–(8).

3. Results and Discussion

The optimization of the function F was carried out with the use of the determined particle size fractions, the type of collector and its dosage. At the second stage, the optimal values were obtained with the use of the determined particle size fraction and the type of collector; finally, it was carried out only with the use of the assumed particle size fraction. The obtained results are shown in Tables 6–8.

Table 6. The optimal values obtained by assumed particle size fraction, collector type, and its dosage.

Assumed Values				Optimal Values		
Particle Size d (μm)	Type of Collector k	Dosage of Collector s (g/t)	t (min)	β	ϑ	ε
0–20	E	100	17	0.108	0.005	0.832
0–20	E	150	17	0.081	0.004	0.839
0–20	I	100	22	0.119	0.005	0.882
0–20	I	150	22	0.103	0.005	0.870
20–40	E	100	3	0.071	0.001	0.677
20–40	E	150	30	0.071	0.013	0.610
20–40	I	100	12	0.063	0.010	0.589
20–40	I	150	12	0.010	0.009	0.706
40–71	E	100	22	0.055	0.011	0.580
40–71	E	150	30	0.062	0.008	0.700
40–71	I	100	17	0.057	0.017	0.446
40–71	I	150	22	0.055	0.019	0.473

Table 7. Optimal values obtained by assumed particle size fraction and type of collector.

Assumed Values			Optimal Values			
Particle Size d (μm)	Type of Collector k	Dosage of Collector s (g/T)	t (min)	β	ϑ	ε
0–20	E	100	22	0.104	0.005	0.858
0–20	I	150	22	0.119	0.005	0.882
20–40	E	100	30	0.071	0.011	0.677
20–40	I	150	12	0.105	0.009	0.709
40–71	E	100	22	0.064	0.009	0.667
40–71	E	150	17	0.057	0.017	0.446

Table 8. Optimal values obtained by the assumed particle size fraction.

Assumed Values	Optimal Values					
Particle Size d (μm)	Type of Collector k	Dosage of Collector s (g/T)	t (min)	β	ϑ	ε
0–20	I	100	22	0.119	0.005	0.882
20–40	I	150	12	0.105	0.009	0.706
40–71	E	150	22	0.064	0.009	0.667

Table 6 shows the calculated indices of optimal values for the sought indices β, ϑ and ε for the assumed particle fractions, the collector type and the dosage. The analysis of the obtained results made it possible to observe that the best quality concentrate, with a copper content amounting to 11.9% for the type 1 collector in the amount of 100 g/t, for the finest particle fraction, within 22 min, was obtained for the finest particle fraction −20 μm. Satisfactory copper recovery in an 88.2% concentrate and copper content in tailings of 0.5% were also obtained in these conditions of the flotation process. For particles of an average size, floating in the Jameson cell 20–40 μm, the taxonomic analysis showed that at a lower dosage of both types of reagents, comparable results −7.1% and 6.3%, respectively, were obtained with regard to β. On the other hand, much better optimal conditions of recovery $\varepsilon = 67.7\%$ and copper content in tailings $\vartheta = 0.1\%$ were obtained for type E reagents in the first three minutes of flotation. Together with an increase in the dosage of type I reagent to 150 g/t, the recovery increases to approx. 70%, but has a negative impact on β and ϑ. In the case of the coarsest floating particle fraction 40–71 μm, the optimal β, ϑ, and ε indices, calculated according to the presented method, in each case reached the lowest values.

Table 7 shows the calculated indices of optimal values for the searched β, ϑ, and ε indices, for the assumed particle fractions and collector types. It is worth noting that with the use of the type 1 collector for the finest particle fraction, better optimal results are obtained with a higher dosage within the same time. Similarly, in the case of a medium size fraction, higher optimal indices were obtained for a higher collector dosage 150 g/t of type 1, β = 10.5%, ϑ = 0.9%, and ε = 70.9% within less than 12 min. For the coarse particle fraction, the optimal β, ϑ, and ε values were obtained for the type E collector, 100 g/t of dosage, but within a longer time.

Table 8 shows indices of optimal values for the sought β, ϑ, and ε indices for the assumed particle fractions. The best optimum rates were obtained for the finest particle fraction with the use of the type I collector, a dosage of 100g/t and during a 22-min flotation.

The next stage was to perform the optimization within the assumed time. The results of this stage are presented in Table 9. If we take into account the flotation type, the best optimum rates were obtained for the finest particle size fraction −20 μm. In this case, the highest β value was determined at the level of 17.5% after 4 min of flotation time. The highest values of the indicators, ϑ = 0.4% and ε = 89.5%, were obtained after 30 min of flotation. Hence, the conclusion is that the longer the time of flotation, the higher is the recovery and the lower the copper content in the waste in the given process conditions for the finest particles.

Table 9. Optimal values obtained by assumed time.

Assumed Values		Optimal Values				
t (min)	Particle Size *d* (μm)	Type of Collector *k*	Dosage of Collector *s* (g/t)	β	ϑ	ε
1	20–40	I	150	0.167	0.021	0.137
2	0–20	E	150	0.121	0.013	0.402
4	0–20	I	100	0.175	0.013	0.613
6	0–20	I	100	0.156	0.009	0.737
9	0–20	I	100	0.142	0.007	0.833
12	0–20	I	100	0.123	0.006	0.856
22	0–20	I	100	0.119	0.005	0.882
30	0–20	I	100	0.112	0.004	0.895

The relations between the optimal values of β, ϑ, ε, and time t are presented in Figures 2–4.

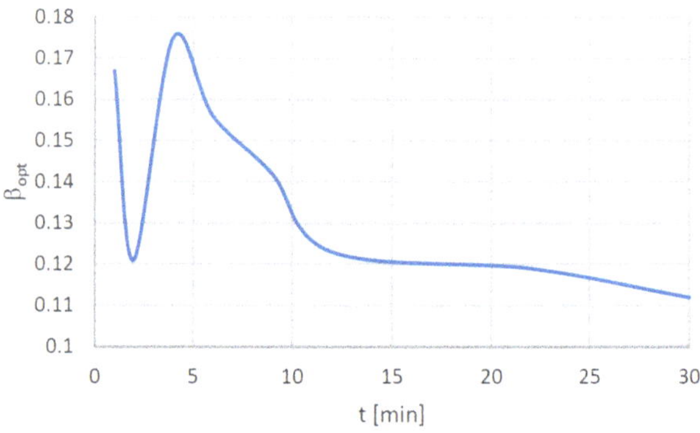

Figure 2. Relation β_{opt}(t).

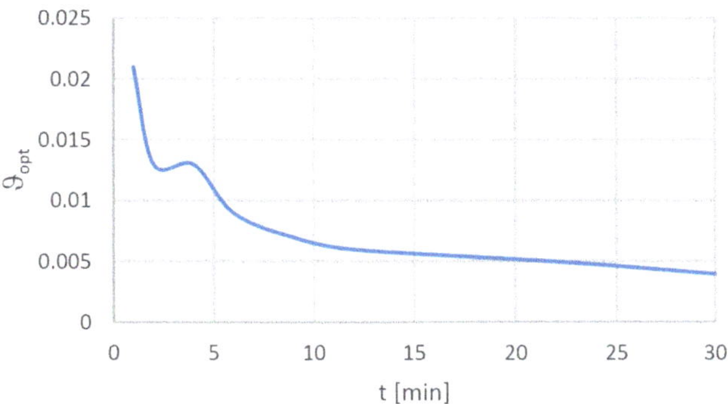

Figure 3. Relation $\vartheta_{opt}(t)$.

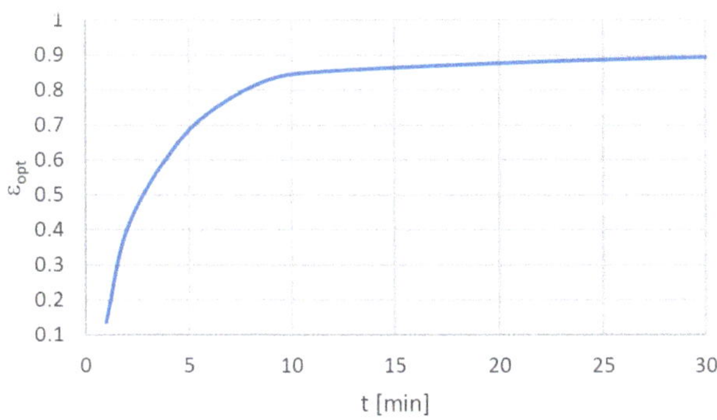

Figure 4. Relation $\varepsilon_{opt}(t)$.

The last stage concerned the optimization in a set of considered values of adjustable variables (Table 2) and the solution is presented in Table 10. Therefore, the optimum conditions of the process were found.

Table 10. The optimal values by values of adjustable variables presented in Table 2.

Particle Size d (µm)	Type of Collector k	Collector Dosage s (g/T)	t (min)	β	ϑ	ε	F_{opt}
0–20	I	100	22	0.119	0.005	0.882	0.376

On the basis of the results, it can be said that the best particle size fraction for the process is 0–20 µm with Aqueous solution of isobutyl sodium xanthate in a dosage of 100 g/T. The optimal time of flotation is 22 minutes. Therefore, the optimal value of the function F is equal to 0.376; it is related to the values of β, ϑ, and ε as 11.9%, 0.5%, and 88.2%, respectively

The Jameson cell has problematic behavior in coarse particle flotation. Sahbaz et al. [40] proved that the maximum size of floating particles having different hydrophobicity degree in various hydrodynamic regions can differ. The results presented in this paper were based on the flotation tests performed in a Jameson cell of the same geometrical properties as was used in [40]. Experiments and literature findings indicate that the turbulence is the most significant parameter in the coarse particle flotation. The stability of the aggregate starts to decrease as the particle size increases, meaning that the detachment force starts to overwhelm the attachment force [38,52–54]. Furthermore, the finer fraction accumulates the biggest amount of copper. This is the reason why this particle size fraction has the biggest potential of copper particle recovery in special conditions. The Jameson cell serves best for this purpose because of its construction and characteristic air bubbles size distribution [55]. The Jameson cell has significant potential to separate fine particle due to very fine bubble production [56,57]. In this test, the results for fine particles are quite good due to these characteristics. In addition, the liberation degree of the sample is higher for the finer size fraction [58]. A conventional cell shows problematic flotation for finer particles due to coarse bubble size causing low collision probability [37,38].

4. Conclusions

The methodology of the optimization of copper flotation results, consisting in the use of the taxonomic method with regard to the beneficiation in a Jameson cell made it possible to determine the optimal conditions of its operation, depending on variable factors, namely the size of particles, the type and dosage of reagent, flotation type for the evaluation indicators of key processes, commonly used in raw material processing. Analyzing the obtained results, it can be observed that for almost all values of time (except $t = 1$ min and $t = 2$ min) the best type of the collector was an Aqueous solution of isobutyl sodium xanthate. It is worth noticing that the best dosage of the collector for the time $t \geq 4$ (min) was a dosage of 100 g/t, while for the time $t < 4$ min it was a dosage of 150 g/t. For smaller particle size fractions (0–20 and 20–40 μm), the Aqueous solution of isobutyl sodium xanthate was a better type of collector, while for the bigger one (40–71 μm) it was xanthate. Analyzing the process depending on particle size, it can be noticed that the best results were obtained definitely for the particle size fraction 0–20. The optimal time in individual cases varied from 12 min to 30 min, but the most suitable time was 22 min. In addition, if the considered indicators are differed in terms of their relevance (if, for example, the economic factors were taken into account), appropriate weights, w_1, w_2, w_3, can be entered into the optimization function. In such a case, particular components of the F function should be multiplied by w_1, w_2, w_3, respectively, where $0 < w_1 < 1, 0 < w_2 < 1$, $0 < w_3 < 1$ and $w_1 + w_2 + w_3 = 1$. The presented methodology can be used efficiently in the evaluation of all kinds of processes and when combined with modeling methods, it can be used as an algorithm of process quality monitoring.

Author Contributions: Conceptualization; methodology; data curation; writing—original draft preparation, T.N.; formal analysis; investigation; writing—original draft, P.P.; writing—review and editing; supervision; validation, A.S., O.Ş. All authors have read and agreed to the published version of the manuscript.

Funding: This research received no external funding.

Acknowledgments: The paper is a result of project no. 11.11.100.276.

Conflicts of Interest: The authors declare no conflict of interest.

References

1. Bulatovic, S.M. *Handbook of Flotation Reagents Chemistry, Theory and Practice: Flotation of Sulfide Ores*; Elsevier Science & Technology: Amsterdam, The Netherlands, 2007.
2. Wang, L.; Xing, Y.; Wang, J. Mechanism of the combined effects of air rate and froth depth on entrainment factor in copper flotation. *Physicochem. Probl. Miner. Process.* **2020**, *56*, 43–53.
3. Rahman, R.M.; Ata, S.; Jameson, G.J. The effect of flotation variables on the recovery of different particle size fractions in the froth and the pulp. *Int. J. Miner. Process.* **2012**, *106*, 70–77. [CrossRef]
4. Hassanzadeh, A.; Hassas, B.V.; Kouachi, S.; Brabcova, Z.; Çelik, M.S. Effect of bubble size and velocity in chalcopyrite flotation. *Colloids Surf. A* **2016**, *498*, 258–267. [CrossRef]
5. Ucurum, M.; Bayat, O. Effects of operating variables on modified flotation parameters in the mineral separation. *Sep. Purif. Technol.* **2007**, *55*, 173–181. [CrossRef]
6. Piestrzyński, A. *Monograph KGHM Polska Miedź S.A.*; Part 2, Geology, 2.19. Litology; CBPM Cuprum: Lubin, Poland, 1996.
7. Dhar, P.; Thornhill, M.; Rao Kota, H. Investigation of Copper Recovery from a New Copper Ore Deposit (Nussir) in Northern Norway: Dithiophosphates and Xanthate-Dithiophosphate Blend as Collectors. *Minerals* **2019**, *9*, 146. [CrossRef]
8. Dhar, P.; Thornhill, M.; Rao Kota, H. Investigation of Copper Recovery from a New Copper Deposit (Nussir) in Northern-Norway: Thionocarbamates and Xanthate-Thionocarbamate Blend as Collectors. *Minerals* **2019**, *9*, 118. [CrossRef]

9. Filip, G.; Podariu, M. Advanced Recovery of Complex Ores using Emulsions of Non-polar Reagents. *Sci. Bull. Ser. D* **2010**, *24*, 53–56.
10. Zhu, R.; Gu, G.; Chen, Z.; Wang, Y.; Song, S. A New Collector for Effectively Increasing Recovery in Copper Oxide Ore-Staged Flotation. *Minerals* **2019**, *9*, 595. [CrossRef]
11. Ziyadanogullari, R.; Aydin, F. A New Application For Flotation Of Oxidized Copper Ore. *J. Miner. Mater. Charact. Eng.* **2005**, *4*, 67–73. [CrossRef]
12. Gutierrez, L.; Betancourt, F.; Uribe, L.; Maldonado, M. Influence of Seawater on the Degree of Entrainment in the Flotation of a Synthetic Copper Ore. *Minerals* **2020**, *10*, 615. [CrossRef]
13. Phiri, T.; Tepa, C.; Nyati, R. Effect of Desliming on Flotation Response of Kansanshi Mixed Copper Ore. *J. Miner. Mater. Charact. Eng.* **2019**, *7*, 193–212. [CrossRef]
14. Podariu, M.; Ilie, P.; Filip, G. Role of Metallic Electrodes in Flotation Activation Phenomena. *Sci. Bull. Ser. D* **2009**, *23*, 121–124.
15. Han, Y.; Zhu, J.; Shen, L.; Zhou, W.; Ling, Y.; Yang, X.; Wang, S.; Dong, Q. Bubble Size Distribution Characteristics of a Jet-Stirring Coupling Flotation Device. *Minerals* **2019**, *9*, 369. [CrossRef]
16. Hassanzadeh, A.; Sajjady, S.A.; Gholami, H.; Amini, S.; Özkan, S.G. An Improvement on Selective Separation by Applying Ultrasound to Rougher and Re-Cleaner Stages of Copper Flotation. *Minerals* **2020**, *10*, 619. [CrossRef]
17. Azizi, A. A study on the modified flotation parameters and selectivity index in copper flotation. *Part. Sci. Technol.* **2017**, *35*, 38–44. [CrossRef]
18. Azizi, A. Optimization of rougher flotation parameters of the Sarcheshmeh copper ore using a statistical technique. *J. Dispers. Sci. Technol.* **2015**, *36*, 1066–1072. [CrossRef]
19. Moimane, T.; Plackowski, C.; Peng, Y. The critical degree of mineral surface oxidation in copper sulphide flotation. *Miner. Eng.* **2020**, *145*, 106075. [CrossRef]
20. Matsuoka, H.; Mitsuhashi, K.; Kawata, M.; Tokoro, C. Derivation of Flotation Kinetic Model for Activated and Depressed Copper Sulfide Minerals. *Minerals* **2020**, *10*, 1027. [CrossRef]
21. Wang, S.; Li, Y.; Zhai, X.; Guan, W. A Recognition Method based on Improved Watershed Segmentation Algorithm or Copper Flotation Conditions. In Proceedings of the 2019 Chinese Automation Congress (CAC), Hangzhou, China, 22–24 November 2019; pp. 224–231.
22. Wang, Z.; He, D.; Li, B. Clustering of Copper Flotation Process Based on the AP-GMM Algorithm. *IEEE Access* **2019**, *7*, 160650–160659. [CrossRef]
23. Ghodrati, S.; Nakhaei, F.; VandGhorbany, O.; Hekmati, M. Modeling and optimization of chemical reagents to improve copper flotation performance using response surface methodology. *Energy Sour. Part A* **2020**, *42*, 1633–1648. [CrossRef]
24. Bahrami, A.; Ghorbani, Y.; Hosseini, M.R.; Kazemi, F.; Abdollahi, M.; Danesh, A. Combined Effect of Operating Parameters on Separation Efficiency and Kinetics of Copper Flotation. *Min. Metall. Explor.* **2019**, *36*, 409–421. [CrossRef]
25. Hassanzadeh, A.; Firouzi, M.; Albijanic, B.; Celik, M.S. A view on determination of particle–bubble encounter using analytical, experimental and numerical methods. *Miner. Eng.* **2018**, *122*, 296–311. [CrossRef]
26. Saramak, D.; Tumidajski, T.; Skorupska, B. Technological and economic strategies for the optimization of Polish electrolytic copper production plants. *Miner. Eng.* **2010**, *23*, 757–764. [CrossRef]
27. Azizi, A.; Masdarian, M.; Hassanzadeh, A.; Bahri, Z.; Niedoba, T.; Surowiak, A. Parametric optimization in rougher flotation performance of a sulfidized mixed copper ore. *Minerals* **2020**, *10*, 660. [CrossRef]
28. Foszcz, D.; Niedoba, T.; Tumidajski, T. A geometric approach to evaluating the results of Polish copper ores beneficiation. *Gospod. Surowcami Min.* **2018**, *34*, 55–66.
29. Jamróz, D.; Niedoba, T.; Pięta, P.; Surowiak, A. The use of neural networks in combination with evolutionary algorithms to optimise the copper flotation enrichment process. *Appl. Sci.* **2020**, *10*, 3119. [CrossRef]
30. Niedoba, T. Determination of partition surface of grained material by means of non-classical approximation methods of distributions functions of particle size and density. *Gospod. Surowcami Min.* **2016**, *32*, 137–154. [CrossRef]
31. Niedoba, T.; Pięta, P. Applications of ANOVA in mineral processing. *Min. Sci.* **2016**, *23*, 43–54.
32. Pięta, P.; Niedoba, T.; Surowiak, A.; Şahbaz, O.; Karagüzel, C.; Canieren, Ö. Studies on Polish copper ore beneficiation in Jameson cell. In *IOP Conference Series, Materials Science and Engineering*, IOP Publishing: Bristol, UK, 2018; Volume 427, p. 012009.
33. Evans, G.M.; Atkinson, B.W.; Jameson, G.J. The Jameson Cell. *Flotat. Sci. Eng.* **1995**, *11*, 331–363.
34. Harbort, G.; Manlapig, E.V.; Debono, S. A discussion of particle collection within the Jameson Cell downcomer. *T. I. Min. Metall. C* **2002**, *307*, C1–C10.
35. Harbort, G.; Debono, S.; Carr, D.; Lawson, V. Jameson Cell Fundamentals—A revised perspective. *Miner. Eng.* **2003**, *16*, 1091–1101. [CrossRef]
36. Mohanty, M.K.; Honaker, R.Q. Performance optimization of Jameson flotation technology for fine coal cleaning. *Miner. Eng.* **1999**, *12*, 367–381. [CrossRef]
37. Gontijo, F.C.; Fornasiero, D.; Ralston, J. The limits of fine and coarse particle flotation. *Can. J. Chem. Eng.* **2017**, *85*, 739–747. [CrossRef]
38. Kowalczuk, P.B.; Şahbaz, O.; Drzymała, J. Maximum size of floating particles in different flotation cells. *Miner. Eng.* **2011**, *24*, 766–771. [CrossRef]

39. Şahbaz, O.; Oteyaka, B.; Kelebek, Ş.; Uçar, A.; Demir, U. Separation of unburned carbonaceous matter in bottom ash using Jameson cell. *Sep. Purif. Technol.* **2008**, *62*, 103–109. [CrossRef]
40. Şahbaz, O.; Uçar, A.; Oteyaka, B. Velocity gradient and maximum floatable particle size in the Jameson cell. *Miner. Eng.* **2013**, *41*, 79–85. [CrossRef]
41. Foszcz, D. *Rules of Determining the Optimal Results of Multi-Component Copper Ores Beneficiation*; IGSMiE PAN: Kraków, Poland, 2013.
42. Wieniewski, A.; Skorupska, B. Technology of Polish copper ore beneficiation—Perspectives from the past experience. In *E3S Web of Conferences*; EDP Sciences: Les Ulis, France, 2016; Volume 8, p. 01064.
43. Aldrich, C. Cluster analysis of mineral process data with autoassociative neural networks. *Chem. Eng. Commun.* **2000**, *177*, 121–137. [CrossRef]
44. Ginsberg, D.W.; Whiten, W.J. The application of clustering to the calibration of onstream analysis equipment. *Int. J. Miner. Process.* **1992**, *36*, 63–79. [CrossRef]
45. Laine, S.; Lappalainen, H.; Jämsä-Jounela, S.L. One-line determination of ore type cluster analysis and neural networks. *Miner. Eng.* **1995**, *6*, 637–648. [CrossRef]
46. Whiteley, J.R.; Davis, J.F. A similarity-based approach to interpretation of sensor data using adaptive resonance theory. *Comput. Chem. Eng.* **1994**, *18*, 637–661. [CrossRef]
47. Ginsberg, D.W.; Whiten, W.J. Cluster analysis for mineral processing applications. *T. I. Min. Metall. C* **1991**, *100*, 139–146.
48. Niedoba, T. Methodological elements of applying two—And multidimensional distributions of grained materials properties to coal beneficiation. *Gospod. Surowcami Min.* **2013**, *29*, 155–172. [CrossRef]
49. Tumidajski, T. Actual tendencies in description and mathematical modeling of mineral processing. *Gospod. Surowcami Min.* **2010**, *26*, 111–123.
50. Nakhaei, F.; Irannajad, M.; Sam, A.; Jamalzadeh, A. Application of d-optimal design for optimizing copper-molybdenum sulphides flotation. *Physicochem. Probl. Miner. Process.* **2015**, *52*, 252–267.
51. Łuniewska, M.; Tarczyński, W. *Methods of Multidimensional Comparative Analysis on the Capital Market*; PWN: Warszawa, Poland, 2006.
52. Drzymała, J. Characterization of materials by Hallimonf tube flotation. *Int. J. Miner. Process.* **1994**, *42*, 139–152. [CrossRef]
53. Schulze, H.J. Dimensionless number and approximate calculation of the upper particle size of floatability in flotation machines. *Int. J. Miner. Process.* **1982**, *9*, 321–328. [CrossRef]
54. Trahar, W.J. A rational interpretation of the role of particle size in flotation. *Int. J. Miner. Process.* **1981**, *8*, 289–327. [CrossRef]
55. Evans, G.M.; Atkinson, B.; Jameson, G.J. The Jameson cell. In *Flotation Science and Engineering. Marcel Dekker*; Matis, K.A., Ed.; Wiley Online Library: Hoboken, NJ, USA, 1995; pp. 331–363.
56. Jameson, G.J. New directions in flotation machine design. *Miner. Eng.* **2010**, *23*, 835–842. [CrossRef]
57. Jameson, G.J.; Goel, S. New approaches to particle attachment and detachment in flotation. In *Separation Technologies for Minerals, Coal, and Earth Resources*; Society for Mining, Metallurgy, and Exploration; Young, C.A., Luttrell, G.H., Eds.; SME: Englewood, CO, USA, 2012; pp. 437–447.
58. Jameson, G.J. The effect of surface liberation and particle size on flotation rate constants. *Miner. Eng.* **2012**, *36–38*, 132–137. [CrossRef]

Article

Kinetic Energy Calculation in Granite Particles Comminution Considering Movement Characteristics and Spatial Distribution

Qing Guo [1,2], Yongtai Pan [1,2,*], Qiang Zhou [1,2], Chuan Zhang [1,2] and Yankun Bi [1,2]

[1] School of Chemical and Environmental Engineering, China University of Mining and Technology (Beijing), Beijing 100083, China; tbp1600301001z@student.cumtb.edu.cn (Q.G.); bqt1700301034@student.cumtb.edu.cn (Q.Z.); bqt2000301006@student.cumtb.edu.cn (C.Z.); byanking945@gmail.com (Y.B.)

[2] Engineering Research Center for Mine and Municipal Solid Waste Recycling, Chemical Engineering and Technology, China University of Mining and Technology (Beijing), Beijing 100083, China

* Correspondence: panyongtai@cumtb.edu.cn; Tel.: +86-15010651331

Abstract: Profound knowledge of the movement characteristics and spatial distribution of the particles under compression during the crushing of rocks and ores is essential to further understanding kinetic energy release law. Various experimental methods such as high-speed camera technology, the coordinate method, and the color tracking method were adopted to improve the understanding of particles' movement characteristics and spatial distribution in rock comminution. The average horizontal velocities of the four size particles α, β, γ, and δ are statistically calculated. The descending order of the particles' average velocity is γ, β, α, and δ. In comparison, the descending order of the particles' kinetic energy is α, β, γ, and δ. Moreover, the contribution of α particles to the total kinetic energy exceeds 70%. The spatial distribution characteristics of coarse and fine particles show different results. The probability of fine particles appearing in the range closer to the center area is greater, while the position of large particles appears to be more random. The color tracking results show that super-large particles generated by crushing are on the specimen's surface, while small particles are generally produced from inside. The above results indicate a connection between the particle generation mechanism, movement characteristics, and spatial distribution in the comminution process.

Keywords: brittle materials; uniaxial compression; comminution; particle size; movement characteristics; particle velocity; kinetic energy; spatial distribution

Citation: Guo, Q.; Pan, Y.; Zhou, Q.; Zhang, C.; Bi, Y. Kinetic Energy Calculation in Granite Particles Comminution Considering Movement Characteristics and Spatial Distribution. *Minerals* **2021**, *11*, 217. https://doi.org/10.3390/min11020217

Academic Editor: Daniel Saramak

Received: 7 January 2021
Accepted: 17 February 2021
Published: 20 February 2021

Publisher's Note: MDPI stays neutral with regard to jurisdictional claims in published maps and institutional affiliations.

1. Introduction

The problem of dynamic fragmentation is a scientific field that has been unresolved for a long time. Compared with the quasi-static fracture of plastic materials, a dynamic fracture is more difficult to understand [1–3]. Dynamic fracture is challenging to study because this process involves complex interactions over an extensive period and space. The main hazard of dynamic fracture is the kinetic energy carried by the ejected fragments during the occurrence. The speed of the destruction of the block sometimes even exceeds 1000 m/s, which is extremely harmful to human activities and the natural environment. [4–6]. The compression and fragmentation of brittle materials are not limited to impact loading. Under the action of the quasi-static compression load, ceramic specimens can still undergo "explosive" damage [7]. Since the research by Mott [8], the dynamic fracture and fragmentation of solids have been a hot research topic. The dynamic fracture of brittle materials can be studied by the uniaxial compression test [9,10], conventional triaxial unloading test [11], true triaxial rock-burst test [12,13], and high-speed impact test [14,15]. Among them, the traditional uniaxial compression and triaxial tests have lower loading rates, which are generally considered to be quasi-static loading, while split Hopkinson pressure bar (SHPB) loading and high-speed impact tests are dynamic loadings [10,16]. Except for conventional triaxial tests restricted by hydraulic cylinders, dynamic fragmentation can be observed in other loading conditions. The most commonly used observation instrument is a high-speed camera that can track particle trajectories and speed measurement [17].

The particle tracking dynamic system can realize the movement tracking of complex and large numbers of particles. This technology is mainly used in high-speed impact tests [18]. The laboratory conducts dynamic fracture experiments of brittle materials to study phenomena such as rock bursts, volcanic eruptions, earthquakes, and planetary collisions. Commonly used experimental materials are basalt [19], quartz [20,21], sandstone, etc. [18]. The research focuses on the particle velocity distribution after dynamic fracture [15], fragment size [10], rebound angle [14], etc.

Energy evolution is a common method for studying dynamic fracture. The quasi-static loading method calculates the input energy through the load-displacement curve [22], and the SHPB loading calculates the absorbed energy of the specimen through the incident and transmitted waves [23]. The high-speed dynamic experiment considers that the kinetic energy of the bullet is input energy [14].

The speed of broken particles can be measured by image tracking technology, and the kinetic energy can be calculated by weighing the particles. Based on the law of conservation of energy, the dissipative heat energy generated by the force-heat coupling process can be studied [24]. Xie [22,25,26] found that studying the energy dissipation and energy release of rock mass structures from the perspective of macroscopic energy conservation can be used to estimate the splash velocity of fragmented rock blocks. Li et al. [10] used SHPB to study the dynamic crushing particle size characteristics, fragment distribution and crushing laws of rock materials. Rait et al. [27] used the discrete element method to study the effect of the loading rate on static fracture and dynamic fracture and analyzed the relationship between the kinetic energy and frictional energy dissipation during the comminution process. Wang et al. [28] studied the energy distribution during the quasi-static confined comminution of granular materials. Xiao et al. [29] analyzed and compared the energy dissipation law of carbonate sand quasi-static and dynamic compression. Zhang [30] studied the average fragmentation and velocity of the debris under a quasi-static load of brittle materials, which agree with the theoretical calculations. The above research mainly focused on the average particle size and velocity and did not involve the velocity and kinetic energy distribution of the characteristic particle size. Exploring the dynamic fracture mechanism of brittle materials requires in-depth research on the speed, kinetic energy, and temporal and spatial distribution characteristics of particles of different sizes produced by crushing.

In response to the above problems, this paper uses high-speed camera technology and digital image motion analysis software to study the velocity–size relationship of particles produced by uniaxial compression crushing of granite and the contribution of products of different sizes to kinetic energy. The coordinate method is used to study the spatial characteristics of fragment distribution at different scales. The color tracking method is used to study the relationship between the spatial characteristics of the fragment distribution and the generation location. The research methods and results have positive significance for describing the splash particles' temporal and spatial characteristics and revealing the kinetic energy release law of the dynamic fracture of brittle materials. At the same time, it is of positive significance for the quantitative calculation of dissipative heat energy and the study of energy evolution in the comminution process.

2. Materials and Methods

2.1. Experimental Materials

The granite was selected from Queshan County, Zhumadian, and all samples were cut and processed from a relatively complete ore body. Firstly, a cylindrical core with a diameter of 50 mm was drilled, and then a cylindrical specimen with a height of 100 mm was cut. A total of 15 granite specimens were prepared in this experiment, as shown in Figure 1. The stone grinder and sandpapers were used to grind both ends of the test piece carefully so that the parallelism of the upper and lower surfaces was within 0.05 mm, and the surface flatness was within 0.02 mm. The samples had good integrity and uniformity, and the average uniaxial compressive strength was 110 MPa. The X-ray fluorescence (XRF) test shows that SiO_2 has the highest content in granite, and the detailed content of other substances is shown in Table 1.

Figure 1. Granite specimens.

Table 1. Granite mineral content.

$SiO_2\%$	$Al_2O_3\%$	$Na_2O\%$	$K_2O\%$	$CaO\%$	$Fe_2O_3\%$	$MgO\%$	$TiO_2\%$
67.75	15.66	4.81	3.84	2.73	2.49	1.41	0.318

2.2. Experimental System

The uniaxial compression test of granite specimens was carried out using the method of force loading. The experimental loading rates were 1, 2, 3, 4, 5 kN/s, with five loading rates and three tests for each loading rate. The unloading process had the same rates as the loading process. This test uses the TAW-3000 hydraulic servo test system (Changchun City Chaoyang Test Instrument CO., LTD., Changchun, China) (as shown in Figure 2a). The testing machine has a portal frame with a stiffness greater than 5 GN/m, which can provide an axial force of 3000 kN and a resolution of 20 N. The resolution of the axial deformation of the specimen is 0.5 μm. The high-speed camera used in this experiment has a shooting frequency of 800 Hz and a shooting area of 400 mm × 500 mm, which is used to record the horizontal velocity of the broken particles' movement. The focal length of the lens used in this experiment was 50 mm.

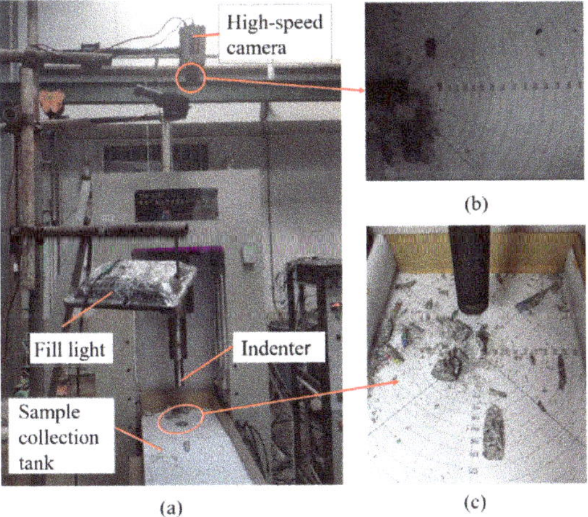

Figure 2. Uniaxial compression crushing dynamic capture system (**a**) Uniaxial compression loading system (**b**) High-speed camera images(**c**) Spatial distribution of fragments.

3. Results and Discussion

3.1. Force-Displacement Relationship of Uniaxial Compression

In the process of the uniaxial compression of the specimen, the displacement of the indenter changed with the load. This change process is usually divided into four stages [31]: the crack compaction stage, elastic stage, microcrack stable-growth stage, and the unstable cracking stage. The accelerated expansion phase and the post-peak segment are shown in Figure 3a. At the same time, energy accumulates, dissipates, and releases inside the specimen. Regardless of the heat exchange between the specimen and the environment, the relationship between input energy, elastic energy, and dissipation energy is as follows [32]:

$$U = U^d + U^e \tag{1}$$

where U is the work done by the external force on the rock, i.e., the energy absorbed by the rock; U^d is the energy dissipated by the rock during the loading process, which is mainly used for the internal damage and plastic deformation of the rock; and U^e is the elastic strain energy stored in the rock. The value of elastic energy can be determined by the area of the unloading curve and the coordinate axis, as shown in Figure 3b. According to the above calculation method, the input energy of specimen 11 before failure is 47.16 J, of which the elastic energy accounts for 24.43 J, with a compression displacement of 0.302 mm.

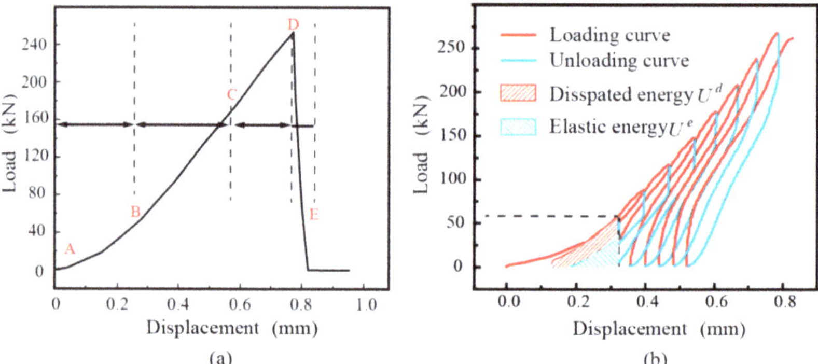

Figure 3. The uniaxial compression load-displacement curve (**a**) Different stages of uniaxial compression (**b**) Calculation of elastic energy and dissipation energy.

3.2. Characteristics of Uniaxial Compression Failure Fragments

In order to facilitate the analysis and study of the movement characteristics of different sizes of uniaxial destruction fragments (at the same time limited by the camera resolution), the fragments obtained after the uniaxial compression experiment were divided into four groups according to the particle size, namely α particles, β particles, γ particles and δ particles [33]. The size of the fragments were divided into +13 mm, 6–13 mm, 3–6 mm, and −3 mm, as shown in Figure 4. Since the fragments were often irregular, the sieving result was used as the measurement and calculation standard during measurement.

The following information can be obtained through observation and analysis of high-speed photography images (Figure 5). In the early stage of macro-destruction, the smaller particles (γ particle) were ejected from the surface of the specimen first. Such particles are located at the front of the detrital cluster and move extremely fast. In the early stage of macro-destruction, the largest particles (α particle) peeled off the surface of the specimen. These particles are located in the front and middle part of the detrital cluster and move faster. In the middle stage of the macro destruction, the larger particles (β particle) peeled off from the surface of the specimen. Such particles are located in the middle of the detrital cluster and move slowly. At the end of macro destruction, the smallest particles (δ particle) were produced, which are located at the back of the detrital cluster and move very slowly. The generation time, spatial location and movement characteristics of the four types of particles were summarized in Table 2.

Figure 4. Classification of fragments produced by uniaxial compression.

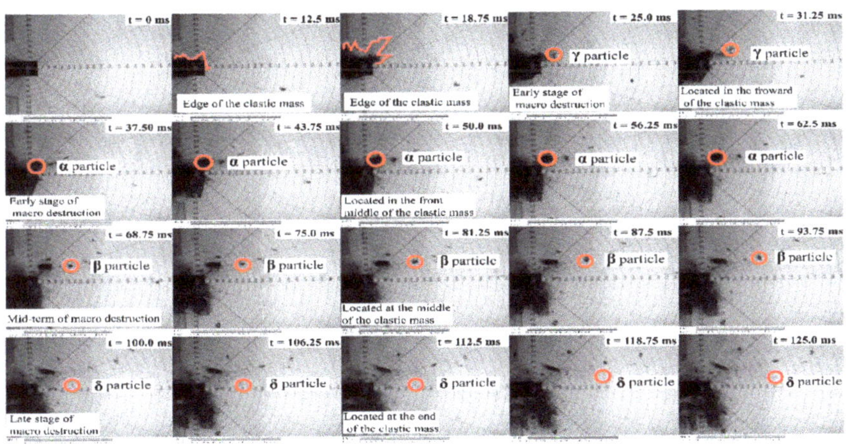

(a)

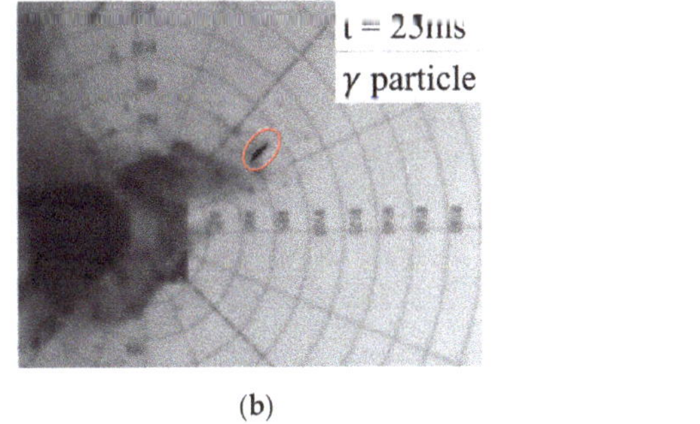

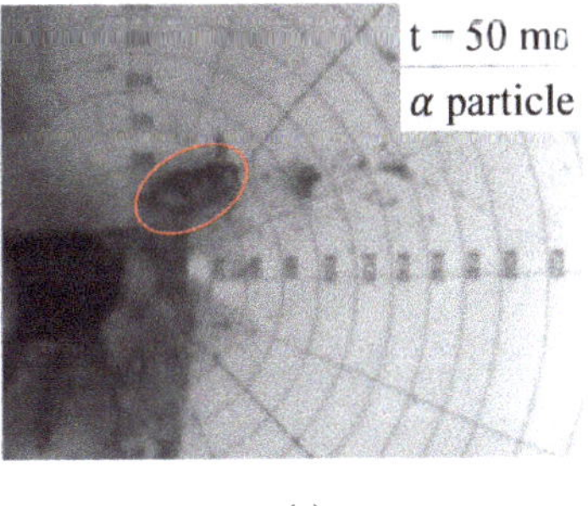

(b)

(c)

Figure 5. *Cont.*

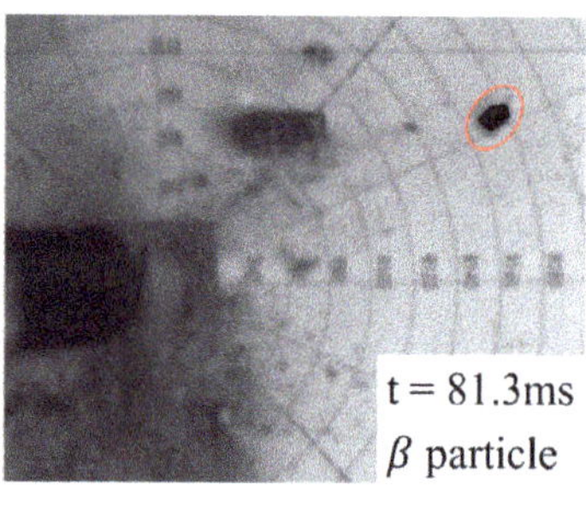

(d)

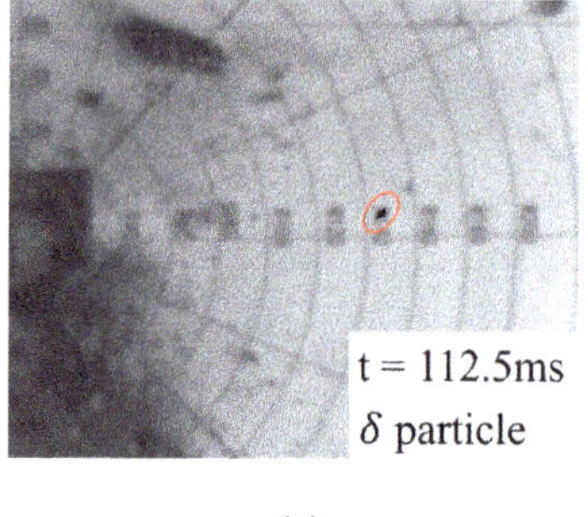

(e)

Figure 5. The temporal and spatial characteristics of the movement of fragments particles. (**a**) High-speed camera images in time series (**b–e**) The enlarged view of the four types of particles.

Table 2. Temporal and spatial position and movement characteristics of fragments particles.

Particle Type	Size/mm	Generation Time	Spatial Location	Movement Characteristics
α	+13	Early stage of macro destruction	Front middle of the detrital cluster	Surface peeling, ejection, Roll along the length
β	6–13	Early and mid-term macro destruction	Middle of detrital cluster	Surface peeling, rotating
γ	3–6	Early stage of macro destruction	Forward of the detrital cluster [34]	Ejection, extremely fast
δ	−3	Mid- to late period of macro destruction	The tail of the detrital cluster	Friction occurs, slower

3.3. Fragments Velocity Characteristics

According to the classification characteristics of Section 3.2, the tracking function of high-speed photography is used to count the horizontal velocity of each sample produced by the representative α, β, γ, and δ particles. In each specimen, about 10 particles were selected as representatives for each of the four particle types. (the super-large particles may be less than 10). The particle size in high-speed photography is measured by the calibration function in the video viewing software. The velocity of the four types of particles in specimen 11 is shown in Table 3. As shown in Table 4, in terms of the average velocity, the descending order is $v_{A\gamma}$, $v_{A\beta}$, $v_{A\alpha}$, and $v_{A\delta}$.

Table 3. Four types of particle velocity of specimen 11.

Serial Number	The Velocity of Particles m/s			
	α Particle	β Particle	γ Particle	δ Particle
1	14.75	8.11	13.28	2.23
2	4.86	8.25	15.08	1.94
3	8.58	6.55	14.97	2.39
4	4.33	8.44	12.15	2.42
5	3.78	8.59	13.09	1.64
6	6.77	6.58	7.63	1.74
7	6.13	7.59	7.71	1.83
8	6.11	7.91	6.63	1.92
9	7.27	6.83	6.40	2.56
10	6.42	7.09	7.90	1.70
v_A	6.900	7.594	10.483	2.036
STD.	2.943	0.740	3.358	0.317

Note: STD. is the abbreviation of standard error values.

Table 4. Four types of particle velocities in different specimens.

Specimen Number	The Average Velocity of Particles m/s			
	α Particle	β Particle	γ Particle	δ Particle
1	3.052	5.827	9.913	1.403
2	4.945	5.151	8.009	2.826
3	2.480	2.825	4.972	1.278
4	2.111	2.104	8.797	1.476
5	4.194	4.052	7.821	3.583
6	2.950	4.864	7.263	2.074
7	4.238	4.856	8.206	2.634
8	2.015	3.604	5.530	1.313
9	3.154	3.160	6.511	1.718
10	6.017	6.357	18.094	2.830
11	6.900	7.594	10.483	2.036
12	5.163	6.834	9.829	2.162
13	6.519	5.434	10.068	2.489
14	2.404	4.109	7.228	1.050
15	2.254	4.909	8.749	1.822
v_A	3.893	4.779	8.765	2.046
STD.	1.615	1.467	2.949	0.691

3.4. Mass Distribution of Fragments

The average value of the horizontal velocity of the four types of particles in 15 groups of specimens is taken as the velocity benchmark for calculating the kinetic energy. The key to calculating kinetic energy is to establish the corresponding relationship between speed and mass. Due to the limited field of view of high-speed photography, it is impossible to match the particles flying on the screen with the particles still in the tray. Therefore, it can only be analyzed by collecting the speed of particles of different characteristic sizes flying through the field of view to form statistical data. The mass of particles with characteristic sizes can be obtained by sieving. Figure 6 shows the sieving data of the four areas—I, II, III, and IV—of specimen 11. The positions of the four zones are shown in Figure 7.

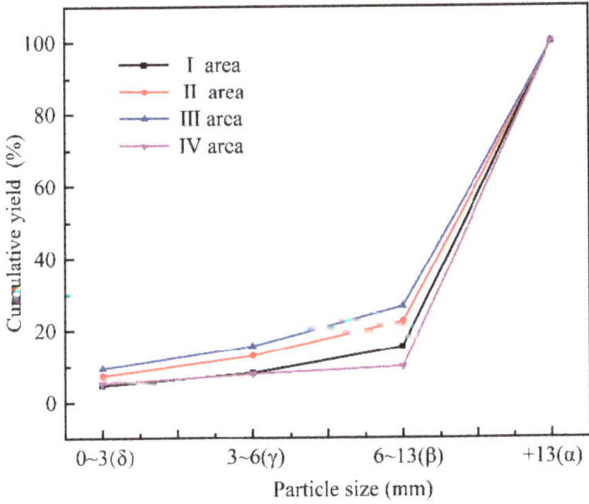

Figure 6. Particle size distribution curve of fragments in different areas of specimen 11.

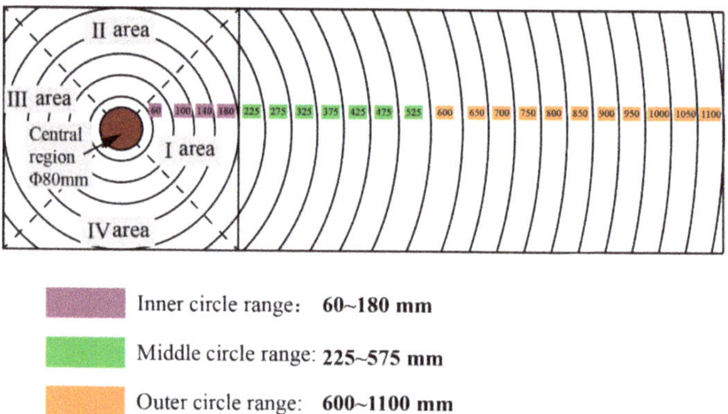

Figure 7. Division of the spatial distribution of fragments.

Theoretically, the distribution of fragments in the four areas after the uniaxial compression failure of homogeneous materials is the same. However, due to the differences in the internal cracks of the materials, the mass distribution of different specimens after crushing is random. Figure 8 shows the proportion of fragments in each area after crushing the five groups of specimens. In most cases, the central area accounts for the largest proportion, with an average mass proportion of 45%. The loading rate variation range of the center area mass between 1–4 kN shows a decreasing trend with the loading rate increase. The mass proportions of the remaining four regions show strong randomness in a single experiment, with an average mass proportion of 10 to 20%. If the four peripheral areas are regarded as a whole, it is opposite to the changing trend of the mass of the central area, and its total mass shows a law of increasing with the increase of loading rate.

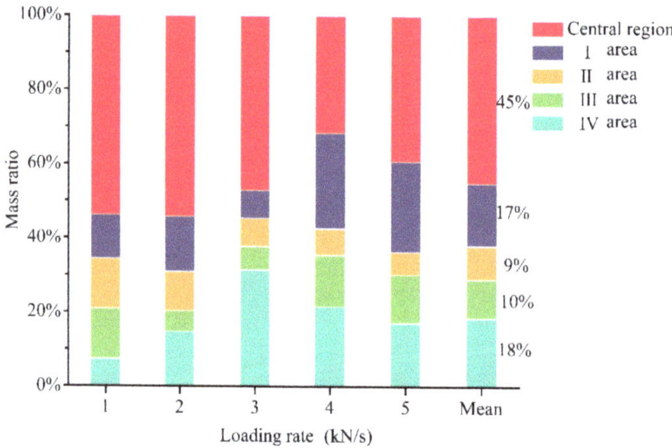

Figure 8. Mass distribution of fragments partition.

3.5. Kinetic Energy of Single-Axis Destruction Fragments

According to the average velocity of the four types of particles and the size distribution of the fragments in each area, the total kinetic energy of each specimen was calculated by using Equation (2):

$$E_k = \frac{1}{2}\left(\sum m_\alpha v_{A\alpha}^2 + \sum m_\beta v_{A\beta}^2 + \sum m_\gamma v_{A\gamma}^2 + \sum m_\delta v_{A\delta}^2\right) \tag{2}$$

where $\sum m_\alpha$ is the sum mass of α particles, $v_{A\alpha}$ is the average velocity values of the α particles. Correspondingly, other symbols indicate the total mass and average velocity of various particles of β, γ, δ.

Table 5 shows the kinetic energy released by each area in the crushing process and the proportion of kinetic energy corresponding to various particles. Sum the total kinetic energy of each area to obtain the total kinetic energy of 5.184 J released by the crushing of specimen 11. Similarly, the kinetic energy released by the crushing of other specimens can be calculated. According to the calculation method described in 3.1, the input energy and elastic energy data of each test piece are calculated, as shown in Table 6. For specimen 11, the ratio of input energy into kinetic energy is 10.79%, and the ratio of elastic energy into kinetic energy is 20.84%. The average of the ratio of kinetic energy to elastic energy of all specimens is 16.03%, and the average of the ratio of kinetic energy to input energy is 7.92%.

Table 5. Fragments kinetic energy in different areas of specimen 11.

Particle Type	I Area		II Area		III Area		IV Area	
	E_k/mJ	PCT.%	E_k/mJ	PCT.%	E_k/mJ	PCT.%	E_k/mJ	PCT.%
α	927.51	73.99	993.57	64.76	818.92	62.35	945.77	87.28
β	167.85	13.39	249.76	16.28	235.87	17.96	38.06	3.51
γ	150.28	11.99	276.82	18.04	244.18	18.59	92.59	8.54
δ	7.94	0.63	14.13	0.92	14.49	1.10	7.15	0.66
Total E_k	1253.58	100	1534.29	100	1313.45	100	1083.57	100

Note: PCT. is the abbreviation of percent.

Table 6. Input energy, elastic energy and kinetic energy of different specimens.

Specimen Number	U	U^e J	E_k mJ	E_k/U^e %	E_k/U %
1	56.21	30.87	1435.36	5.21	2.55
2	40.53	23.09	2414.18	13.36	5.96
3	25.39	12.53	626.53	5.85	2.47
4	33.78	15.88	1073.90	6.21	3.18
5	27.65	12.36	2281.99	18.50	8.25
6	44.28	25.69	1441.39	6.11	3.25
7	23.72	8.38	3481.25	30.44	14.68
8	25.98	14.05	823.46	5.98	3.17
9	26.09	10.62	1793.53	16.06	6.88
10	41.20	20.98	4369.55	25.84	10.61
11	47.16	24.43	5184.90	21.22	10.99
12	21.01	8.56	2542.20	30.55	12.10
13	27.90	11.87	7289.66	54.95	26.13
14	30.01	13.75	1557.10	9.01	5.19
15	41.37	20.81	1383.43	7.87	3.34
Average	34.15	16.25	2604.32	16.03%	7.92%

As there are few studies on the kinetic energy calculation of the rock fragmentation under uniaxial compression, the author has not found convincing data to verify it. However, in similar destruction modes, the proportion of kinetic energy can be used as evidence. For example, the impact of spherical particles [35], rock blasting [36] and the ratio of kinetic energy to input energy are approximately 3% and 3–21%, respectively. In the true triaxial failure of brittle rocks [37], the ratio of kinetic energy to elastic energy ranges from 8 to 50%. In dynamic fracture of pre-cracked rock specimens, the SHPB system was used, and the ratio of kinetic energy to input energy ranges from 22 to 59% [38].

The kinetic energy of the four kinds of particles generated by the crushing of the specimen at different loading rates is shown in Figure 9. It can be seen that the total kinetic energy increases with the increase of the loading rate within the range of loading rate of 1–4 kN/s. Furthermore, the main factor affecting the total kinetic energy is the kinetic energy of α particles. The relationship between the kinetic energy of other types of particles and the loading rate is not obvious. When the loading rate is 5 kN/s, the total kinetic energy decreases, which is mainly affected by the decrease of the kinetic energy of α particles.

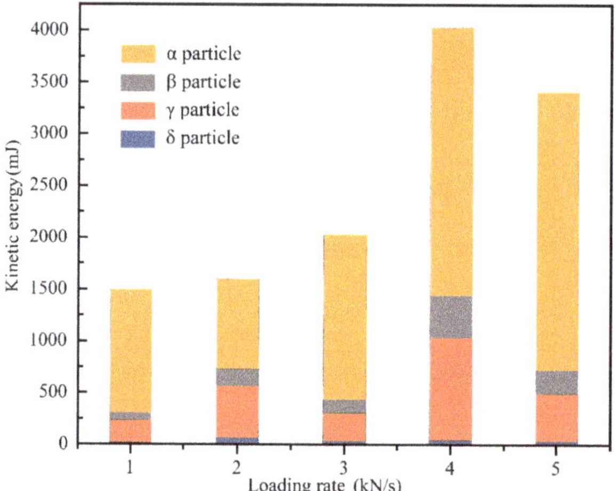

Figure 9. The kinetic energy of four particles under different loading rates.

The proportion of kinetic energy of various particles in specimen 11 can be obtained by summarizing the data in Table 5. The distribution law of kinetic energy can be seen in Figure 10; the kinetic energy proportions of the four types of particles are ranked from high to low as $E_{K\alpha} > E_{K\gamma} > E_{K\beta} > E_{K\delta}$. The kinetic energy of α particles accounts for about 70%, the kinetic energy of γ particles accounts for close to 20%, the kinetic energy of β particles accounts for close to 10%, and the kinetic energy of δ particles accounts for about 1.5%. Comparing the average values of specimen 11 and specimens 1 to 15 shows that the kinetic energy distribution of various particles of a single specimen is not significantly different from the overall distribution. The two indicators that affect the magnitude of kinetic energy are speed and quality. The α-type particles have the largest mass, and the γ particles have the largest velocity. Since the mass of alpha particles is more than an order of magnitude higher than that of gamma particles, the speed of γ particles is several times that of alpha particles. This has led to massive particles becoming the main contributor to kinetic energy.

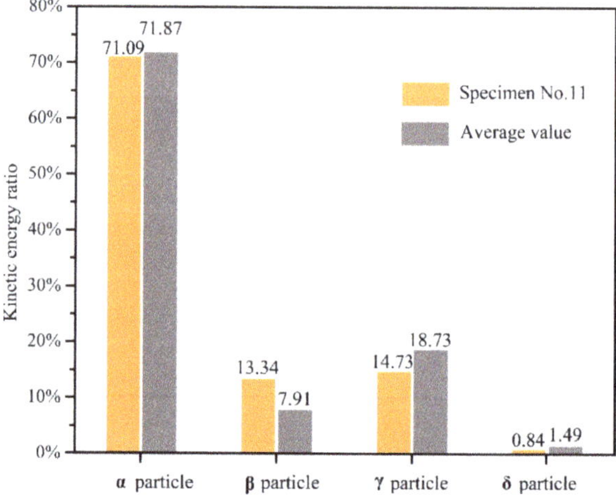

Figure 10. Percentage of the kinetic energy of various particles.

3.6. Spatial Distribution of Fragments

With a diameter of 6 mm as the standard, the fragments were divided into large particles and fine particles. The masses of the fragments in the I area and the extended area along the radial direction were counted after crushing. Figure 11a shows the spatial distribution of fine particles generated by the crushing of as there are few studies on the all specimens. The density of the data points represents the possibility of corresponding mass fine particles in the corresponding area. The blank area near the center area indicates that the mass of fine particles produced in the inner circle is more considerable, generally above 0.1 g. The closer to the center area, the more fine particles. Figure 11b shows the changing trend of the total mass of fine particles of specimens 1 to 15 in the range of 60 to 200 mm, which confirms this rule. Figure 11b shows that the particle mass has a maximum value at 300 mm. This aggregation phenomenon reflects the fine particle velocity distribution characteristics, which represents the intersection of the γ and δ particles. After the maximum point, the mass of fine particles decreases as the distance increases, and the decreasing trend gradually slows down. In the area larger than 1000 mm, the particle mass tends to increase again, mainly because the collection trough restricts fragment movement. The loading rate has no significant effect on the spatial distribution of fine particles.

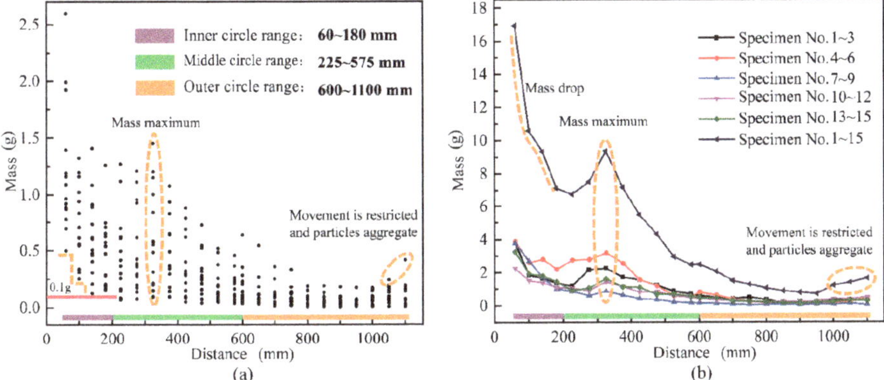

Figure 11. Spatial distribution characteristics of fine particles. (**a**) Scatter plot of particle spatial distribution (**b**) Summary of particle spatial distribution of each group.

The spatial distribution of large particles shows strong randomness, and the probability of super large particles is small. It can be seen from Figure 12a that there are only four particles larger than 70 g in all of the data, but they contribute most of the mass of the inner circle and the middle circle. Among more than 300 sets of data, there were only 16 sets of super-large particles with a mass greater than 20 g. There were only four groups of super large particles in the outer circle, and the mass was less than 50 g. From the spatial distribution of large particles, it can be seen that the input energy of this uniaxial compression and crushing is limited, which is not enough to push the super large particles to a more distant area. The input energy may be related to the material properties and loading rate, and it is worthy of further exploration. For 15 sets of experiments, larger loading rates are more likely to produce large splashing particles. As shown in Figure 12b, the mass of large particles produced by specimens 10–15 (loading rate 4–5 kN/s) is larger than that of specimens 1–9 (loading rate 1–3 kN/s).

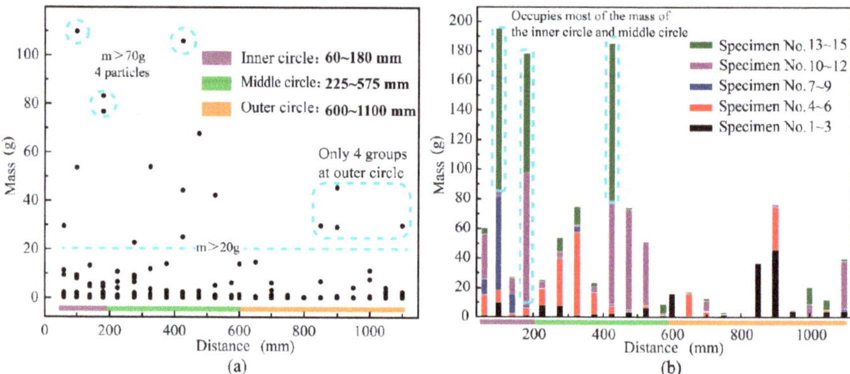

Figure 12. Spatial distribution characteristics of large particles. (**a**) Scatter plot of particle spatial distribution (**b**) Summary of particle spatial distribution of each group.

3.7. Location of Fragments

The surface of the specimen was painted. The debris larger than 6 mm can be divided into surface particles and internal particles according to whether there is a color on the surface. The mass of the two types of particles in area I along the radial direction is counted. On the whole, there is no apparent difference between Figures 12 and 13, which indicates that most of the particles larger than 6mm are surface particles; that is, at least one surface is the surface of the test specimen. On the contrary, it is easier to compare the difference between Figures 12 and 13 from the spatial mass distribution of the internal particles. That is to say, the mass of the particles at each distance in Figure 14b is the difference between the mass of the particles at the corresponding distance in Figures 12 and 13. Therefore, although Figures 12b and 13b are relatively close in morphology, there are differences in the number and mass of particles.

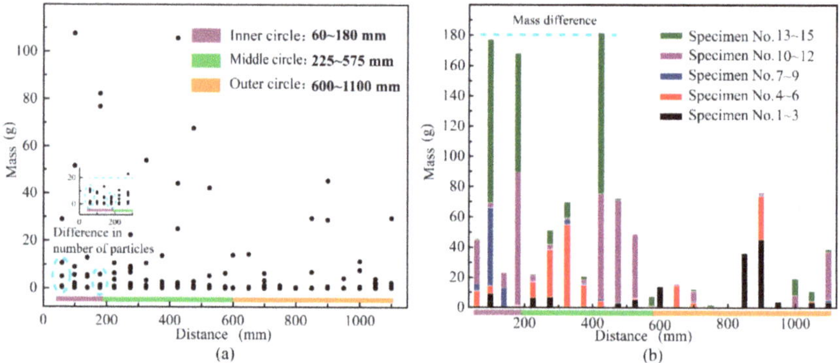

Figure 13. Spatial distribution characteristics of surface particles. (**a**) Scatter plot of particle spatial distribution (**b**) Summary of particle spatial distribution of each group

This part of the difference represents the particles generated from the inside of the specimen. From the mass distribution, it can be found that the mass of the internal particles in this part is small, and the distribution characteristics are similar to the mass-spatial distribution of fine particles in Figure 11. The mass spatial distribution of internal particles presents the following law as a whole: the mass near the center is large, the mass decreases rapidly as the distance increases, and the rate of decrease gradually decreases. Simultaneously, such particles' appearance will still show a certain degree of randomness, and there may even be no particles in some areas. All in all, the spatial distribution characteristics of particles larger than 6 mm generated inside have a part of the characteristics corresponding to particles smaller than 3 mm and particles larger than 6 mm on the surface of the specimen, which belong to the transition type between the two.

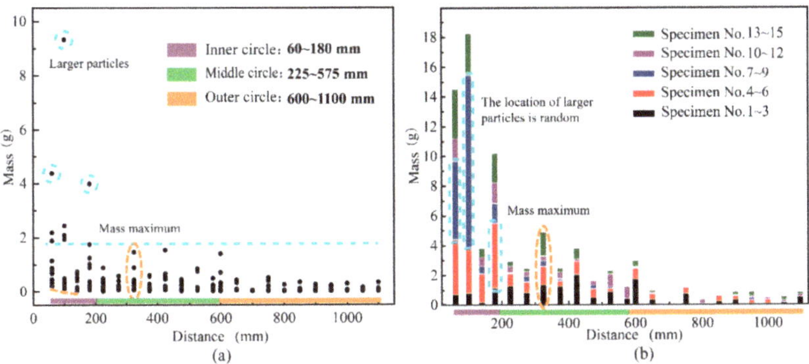

Figure 14. Spatial distribution characteristics of internal particles. (**a**) Scatter plot of particle spatial distribution (**b**) Summary of particle spatial distribution of each group

4. Conclusions

The fragments are divided into four types of particles according to the particle size.

The average horizontal velocities of the four size particles α, β, γ, and δ are statistically calculated. The descending order of the particles' average velocity is γ, β, α, and δ. Since the mass difference of different types of particles is greater than the influence of the velocity difference on kinetic energy, the descending order of the particles' kinetic energy is α, β, γ, and δ. Among them, the contribution of alpha particles to the total kinetic energy exceeds 70%. The loading rate has little effect on the particle velocity. When the loading rate is higher, more alpha particles leave the central area, resulting in more input energy being converted into kinetic energy. The percentage of input energy converted into kinetic energy of specimen 11 is 5.9% during the crushing process.

The spatial distribution characteristics of large particles and fine particles were analyzed by the coordinate method. As a result, it was found that there was a greater probability of fine particles appearing in the range closer to the central area; this reflects that most of the fine particles have a lower velocity. The maximum value of the fine particles' mass appears in the middle circle, which indicates that there are also particles with higher speed in the fine particles, namely γ particles. These kinds of particles overlap with the slower particles, causing the phenomenon of mass maximum. The locations of large particles are random, but they are more likely to appear within the middle circle. A larger loading rate can produce more large splashing particles, which is consistent with the kinetic energy characteristics of the loading rate.

The color tracking method was used to study the location of particles larger than 6 mm in the specimen. It was found that at least one surface of the super large particles produced by crushing was the surface of the test specimen. Those particles produced entirely from the inside of the specimen are relatively small and have similar spatial distribution characteristics to fine particles. Therefore, it can be judged that fine particles and particles of smaller size are generally generated by friction between the cross-sections of the specimen when the specimen is broken. The speed of such particles is generally low. Most of the large particles and a few small particles are directly peeled off the surface of the broken specimen and have a higher splash speed.

Author Contributions: Y.P. provides overall experimental ideas and methods; Q.G. collected and archived the experiment results; Q.G. analyzed the experimental data and initiated the writing of the paper; Q.Z., C.Z. and Y.B. help analyze the experimental data. All authors have read and agreed to the published version of the manuscript.

Funding: The financial support from National Natural Science Foundation of China (Grant No. 52074308).

Conflicts of Interest: The authors declare no conflict of interest.

References

1. Griffith, A.A. The Phenomena of Rupture and Flow in Solids. *Philos. Trans. R. Soc. Lond. Ser. A Contain. Pap. Math. Phys. Character* **1921**, *221*, 98–163. [CrossRef]
2. Rice, J.R. Thermodynamics of the quasi-static growth of Griffith cracks. *J. Mech. Phys. Solids* **1978**, *26*, 61–78. [CrossRef]
3. Irwin, G.R. *Fracture Dynamics: Fracturing of Metals*; American Society for Metals: Cleveland, OH, USA, 1948; pp. 147–166.
4. Taddeucci, J.; Alatorre-Ibargueengoitia, M.A.; Cruz-Vazquez, O.; Del Bello, E.; Scarlato, P.; Ricci, T. In-flight dynamics of volcanic ballistic projectiles. *Rev. Geophys.* **2017**, *55*, 675–718. [CrossRef]
5. Zhao, T.; Crosta, G.B.; Dattola, G.; Utili, S. Dynamic Fragmentation of Jointed Rock Blocks during Rockslide-Avalanches: Insights from Discrete Element Analyses. *J. Geophys. Res. Solid Earth* **2018**, *123*, 3250–3269. [CrossRef]
6. Kim, E.; Garcia, A.; Changani, H. Fragmentation and energy absorption characteristics of Red, Berea and Buff sandstones based on different loading rates and water contents. *Geomech. Eng.* **2018**, *14*, 151–159. [CrossRef]
7. Brannon, R.M.; Lee, M.Y.; Bronowski, D.R. *Uniaxial and Triaxial Compression Tests of Silicon Carbide Ceramics under Quasi-Static Loading Condition*; Sandia National Laboratories: Albuquerque, NM, USA, 2005; pp. 9–26.
8. Mott, N.F. Fragmentation of shell cases. *Proc. R. Soc. Lond. Ser. A Math. Phys. Sci.* **1947**, *189*, 300–308. [CrossRef]
9. Fondriest, M.; Doan, M.-L.; Aben, F.; Fusseis, F.; Mitchell, T.M.; Voorn, M.; Secco, M.; Di Toro, G. Static versus dynamic fracturing in shallow carbonate fault zones. *Earth Planet. Sci. Lett.* **2017**, *461*, 8–19. [CrossRef]
10. Li, X.F.; Li, H.B.; Zhang, Q.B.; Jiang, J.L.; Zhao, J. Dynamic fragmentation of rock material: Characteristic size, fragment distribution and pulverization law. *Eng. Fract. Mech.* **2018**, *199*, 739–759. [CrossRef]
11. Songlin, X.; Wen, W.; Hua, Z. Experimental study on dynamic unloading of the confining pressures for a marble under triaxial compression and simulation analyses of rock burst. *J. Liaoning Tech. Univ.* **2002**, *21*, 612–615. [CrossRef]
12. Manchao, H.; Jinli, M.; Dejiang, L.; Chunguang, W. Experimental study on rockburst processes of granite specimen at great depth. *Chin. J. Roc. Mech. Eng.* **2007**, *26*, 865–876. [CrossRef]
13. Guoshao, S.; Jianqing, J.; Xiating, F.; Chun, M.; Quan, J. Experimental study of ejection process in rockburst. *Chin. J. Roc. Mech. Eng.* **2016**, *35*, 1990–1999. [CrossRef]
14. Hogan, J.D.; Spray, J.G.; Rogers, R.J.; Vincent, G.; Schneider, M. Dynamic fragmentation of natural ceramic tiles: Ejecta measurements and kinetic consequences. *Int. J. Impact Eng.* **2013**, *58*, 1–16. [CrossRef]
15. Hogan, J.D.; Rogers, R.J.; Spray, J.G.; Vincent, G.; Schneider, M. Debris Field Kinetics during the Dynamic Fragmentation of Polyphase Natural Ceramic Blocks. *Exp. Mech.* **2014**, *54*, 211–228. [CrossRef]
16. Gong, D.; Nadolski, S.; Sun, C.; Klein, B.; Kou, J. The effect of strain rate on particle breakage characteristics. *Powder Technol.* **2018**, *339*, 595–605. [CrossRef]
17. Jiang, J.; Su, G.; Zhang, X.; Feng, X.-T. Effect of initial damage on remotely triggered rockburst in granite: An experimental study. *Bull. Eng. Geol. Environ.* **2020**, *79*, 3175–3194. [CrossRef]
18. Hermalyn, B.; Schultz, P.H. Early-stage ejecta velocity distribution for vertical hypervelocity impacts into sand. *Icarus* **2010**, *209*, 866–870. [CrossRef]
19. Fujiwara, A.; Tsukamoto, A. Experimental study on the velocity of fragments in collisional breakup. *Icarus* **1980**, *44*, 142–153. [CrossRef]
20. Hogan, J.D.; Spray, J.G.; Rogers, R.J.; Boonsue, S.; Vincent, G.; Schneider, M. Micro-scale energy dissipation mechanisms during dynamic fracture in natural polyphase ceramic blocks. *Int. J. Impact Eng.* **2011**, *38*, 931–939. [CrossRef]
21. Kimberley, J.; Ramesh, K.T.; Barnouin, O.S. Visualization of the failure of quartz under quasi-static and dynamic compression. *J. Geophys. Res. Solid Earth* **2010**, *115*. [CrossRef]
22. Heping, X.; Yang, J.; Liyun, L.; Ruidong, P. Energy Mechanism of Deformation and Failure of Rock Masses. *Chin. J. Rock Mech. Eng.* **2008**, *27*, 1729–1739. [CrossRef]
23. Zhou, Z.; Cai, X.; Li, X.; Cao, W.; Du, X. Dynamic Response and Energy Evolution of Sandstone Under Coupled Static-Dynamic Compression: Insights from Experimental Study into Deep Rock Engineering Applications. *Rock Mech. Rock Eng.* **2020**, *53*, 1305–1331. [CrossRef]
24. Minh Phono, L. Infrared thermovision of damage processes in concrete and rock. *Eng. Fract. Mech.* **1990**, *35*, 291–301. [CrossRef]
25. Heping, X.; Yang, J.; Liyun, L. Criteria for Strength and Structural Failure of Rocks Based on Energy Dissipation and Energy Release Principles. *Chin. J. Rock Mech. Eng.* **2005**, *24*, 3003–3010.
26. Li, L.y.; Ju, Y.; Zhao, Z.w.; Wang, L.; Lu, J.; Ma, X. Energy analysis of rock structure under static and dyanmic loading conditions. *J. China Coal Soc.* **2009**, *34*, 737–740. [CrossRef]

27. Rait, K.L.; Bowman, E.T.; Lambert, C. Dynamic fragmentation of rock clasts under normal compression in sturzstrom. *Geotech. Lett.* **2012**, *2*, 167–172. [CrossRef]
28. Wang, P.; Arson, C. Energy distribution during the quasi-static confined comminution of granular materials. *Acta Geotech.* **2018**, *13*, 1075–1083. [CrossRef]
29. Xiao, Y.; Yuan, Z.x.; Chu, J.; Liu, H.l.; Huang, J.y.; Luo, S.N.; Wang, S.; Lin, J. Particle breakage and energy dissipation of carbonate sands under quasi-static and dynamic compression. *Acta Geotech.* **2019**, *14*, 1741–1755. [CrossRef]
30. Zhang, Q.; Zheng, Y.; Zhou, F.; Yu, T. Fragmentations of Alumina (Al_2O_3) and Silicon Carbide (SiC) under quasi-static compression. *Int. J. Mech. Sci.* **2020**, *167*, 105119. [CrossRef]
31. Martin, C.D.; Chandler, N.A. The progressive fracture of Lac du Bonnet granite. *Int. J. Rock Mech. Min. Sci. Geomech. Abstr.* **1994**, *31*, 643–659. [CrossRef]
32. Huang, D.; Huang, R.Q.; Zhang, Y.X. Experimental investigations on static loading rate effects on mechanical properties and energy mechanism of coarse crystal grain marble under uniaxial compression. *Chin. J. Rock Mech. Eng.* **2012**, *31*, 245–255. [CrossRef]
33. He, M.; Yang, G.; Miao, J.; Jia, X.; Jiang, T. Classification and research methods of rockburst experimental fragments. *Chin. J. Rock Mech. Eng.* **2009**, *28*, 1521–1529.
34. Bowman, E.T.; Take, W.A.; Rait, K.L.; Hann, C. Physical models of rock avalanche spreading behaviour with dynamic fragmentation. *Can. Geotech. J.* **2012**, *49*, 460–476. [CrossRef]
35. Wu, S.Z.; Chau, K.T.; Yu, T.X. Crushing and fragmentation of brittle spheres under double impact test. *Powder Technol.* **2004**, *143–144*, 41–55. [CrossRef]
36. Sanchidrian, J.A.; Segarra, P.; Lopez, L.M. Energy components in rock blasting. *Int. J. Rock Mech. Min.* **2007**, *44*, 130–147. [CrossRef]
37. Akdag, S.; Karakus, M.; Taheri, A.; Nguyen, G.; Manchao, H. Effects of Thermal Damage on Strain Burst Mechanism for Brittle Rocks Under True-Triaxial Loading Conditions. *Rock Mech. Rock Eng.* **2018**, *51*, 1657–1682. [CrossRef]
38. Sun, Y.; Qi, C.; Zhu, H.; Guo, Y.; Wang, Y. Energy Analysis on Rock Dynamic Fracture Process. *Chin. J. Undergr. Sp. Eng.* **2020**, *16*, 43–49.

Article

A New Belt Ore Image Segmentation Method Based on the Convolutional Neural Network and the Image-Processing Technology

Xiqi Ma [1,2], **Pengyu Zhang** [1,2], **Xiaofei Man** [3] **and Leming Ou** [1,2,*]

1   School of Minerals Processing and Bioengineering, Central South University, Changsha 410083, China; maxiqi@csu.edu.cn (X.M.); pengyu7765@csu.edu.cn (P.Z.)

2   Key Laboratory of Hunan Province for Clean and Efficient Utilization of Strategic Calcium-Containing Mineral Resources, Central South University, Changsha 410083, China

3   Ansteel Beijing Research Institute Co., Ltd., Beijing 102200, China; maryman@csu.edu.cn

*   Correspondence: olmpaper@csu.edu.cn; Tel.: +86-0731-8883-0913

Received: 7 November 2020; Accepted: 9 December 2020; Published: 11 December 2020

Abstract: In the field of mineral processing, an accurate image segmentation method is crucial for measuring the size distribution of run-of-mine ore on the conveyor belts in real time0The image-based measurement is considered to be real time, on-line, inexpensive, and non-intrusive. In this paper, a new belt ore image segmentation method was proposed based on a convolutional neural network and image processing technology. It consisted of a classification model and two segmentation algorithms. A total of 2880 images were collected as an original dataset from the process control system (PCS). The test images were processed using the proposed method, the PCS system, the coarse image segmentation (CIS) algorithm, and the fine image segmentation (FIS) algorithm, respectively. The segmentation results of each algorithm were compared with those of the manual segmentation. All empty belt images in the test images were accurately identified by our method. The maximum error between the segmentation results of our method and the results of manual segmentation is 5.61%. The proposed method can accurately identify the empty belt images and segment the coarse material images and mixed material images with high accuracy. Notably, it can be used as a brand new algorithm for belt ore image processing.

Keywords: belt ore measurement; convolutional neural network; image processing; contour detection; OpenCV

1. Introduction

The particle size distribution of run-of-mine ore exhibits a great influence on the grinding process. Variations in the particle size distribution directly affect the throughput and power consumption of mills, especially autogenous (AG) and semi-autogenous (SAG) grinding mills [1]. Therefore, it is critical to evaluate the size distribution of run-of-mine ore on the conveyor belts in real time [2,3]. The measurement of the particle size distribution by sampling and sieving is considered a common and time-consuming method. The analysis method based on machine vision is considered a non-invasive, fast, and inexpensive technique for rock size measurement [4]. Since the 1980s, many studies have been conducted to evaluate the particle size distribution of materials on a conveyor belt based on machine vision and image processing technology [2,5,6]. Scholars have mainly followed three aspects of exploration. The first aspect includes accurate ore contour detection algorithms. The second aspect is the reasonable evaluation model which is used to convert two-dimensional information of ore into three-dimensional information, and then evaluating energy consumption or particle size distribution. The third aspect is new technology including neural networks, deep learning, and genetic algorithms, etc. In 1988, Lange developed an on-line, real-time system which can capture images of rocks on conveyor belts, and process images to get the chord-length distributions, and then transform chord-length distributions to equivalent sieve sizes. Lange offered a method to distinguish belt ores of different size distributions [2]. Lin and Miller developed an image-based system which used image processing technology to get the chord-length of rocks, and used two kernel functions to calculate the cumulative chord-length distributions of regularly and irregular shaped particles, respectively. The last step was to transform the chord-length distributions into size distributions by the transformation Equation [5]. In 1997, Yen and co-workers used an empirical correction function to solve the coarse particle overlap problem [7]. Before 2000, limited by hardware technology, it was difficult to get sharp images and many algorithms that consumed too much computer performance could not be adopted. The scholars mainly researched the image-based system software and hardware framework, contour detection, reasonable measurement parameters and size transformation functions.

After 2000, with the rapid development of computer hardware and new technologies, the image-based, online, and real-time particle size measurement development made much progress. Singh and Mohan Rao extracted RGB color information, visual texture of particles, and developed a system based on a radial basis neural network. The system was used for ore classification and ore sorting [8]. Al-Thyabat and co-workers evaluated ore image segmentation results by means of Feret's diameter and equivalent area diameter, and experimented and discussed the effect of camera positions [9]. Levner offered a classification-driven watershed segmentation to segment belt ore images, and adopted machine learning to produce markers and identify ore edges [10]. Outal et al. provided a calibration method for evaluating 3D size distribution, according to the 2D segmentation results [11]. Andersson evaluated the size distribution of particles using ordinal logistic regression [12]. Hamzeloo et al. used different particle equivalent models to evaluate 3D size distribution. These equivalent models included the equivalent area circle, best-fit rectangle, Feret diameter, and maximum inscribed disk [4]. In addition, the impact of the shape of the particles on the product properties was also researched extensively [13–16]. Until now, many image-based analysis methods have been successfully applied to evaluate the particle size distribution [17–19], however there are still three unresolved problems regarding the image-based, on-line particle size analysis methods.

One is that there is no research focus on the problem of empty belt identification. In the course of production, we should not turn on or switch to the empty conveyor belts. Therefore, the accurate recognition of the empty belt is necessary to realize the automatic control and switching of conveyor belts. The second problem is the accurate belt ore image segmentation method, especially for the coarse-fine images. According to the experimental results of previous researches, it cannot be concluded that the proposed method can accurately segment both coarse and fine materials. An accurate image segmentation method is the basis of a particle size distribution measurement system. Although there is a method which uses machine learning to identify the pixels of ore edges, the contour detection based on image segmentation is more stable, accurate, and adaptable. In the future, the ore image segmentation method based on deep learning will be an expected practice. The third problem is related to the overlapping of particles. Even though many studies have been conducted in order to find viable solutions to these problems, most of them rely on empirical correction [4,7,20]. Dynamic image analysis (DIA) is a feasible method to solve the overlap problem [15].

Traditional rock image analysis methods cannot distinguish different types of images, such as empty belt images, mixed material images, and coarse material images, which are distinct. The mixed materials include coarse-fine materials and fine materials. It is difficult to accurately process all three types of belt ore images with an image segmentation algorithm; therefore, our analysis method should be able to accurately classify the images we obtain. In recent years, with the rapid development of computer hardware and deep learning theory, the convolutional neural networks (CNNs) have shown great progress in the field of image recognition and classification [21,22]. Krizhevsky et al. developed AlexNet, which can reach 83.6% top-5 accuracy for the ImageNet dataset [23]. At present, the top-5 accuracy of many convolutional neural networks in image recognition tasks can reach more than 90%

for the ImageNet dataset [24–26]. Many research studies on rock image recognition and classification based on deep learning have achieved high accuracy [8,27,28].

In this research, the method based on the deep learning method and the image processing technology was developed in order to accurately segment the belt ore images. The strategy was to classify the belt ore images into empty belt, mixed materials and coarse materials first and then use different algorithms for processing mixed materials and coarse materials. We focused on the accuracy of belt ore image segmentation and empty belt identification. Both the conversion model for converting two-dimensional information of ore into three-dimensional information and the impact of the shape of the particles on the product properties are beyond the scope of this article.

2. Details of the Method

The proposed method is divided into three layers. The first layer is a classifier based on a convolutional neural network. The second layer consists of two image processing algorithms based on the OpenCV library. The two algorithms are used to process coarse material images and mixed material images, respectively. The third layer is the statistics layer. The classifier divides the raw images into the empty belt, coarse materials, and mixed materials. If the belt is empty, it gives an alarm; otherwise, it uses the coarse image segmentation (CIS) algorithm to process coarse material images and uses the fine image segmentation (FIS) algorithm to process mixed material images, respectively. Finally, the cumulative area distribution is calculated following the counting segmentation area information. The technical roadmap is shown in Figure 1.

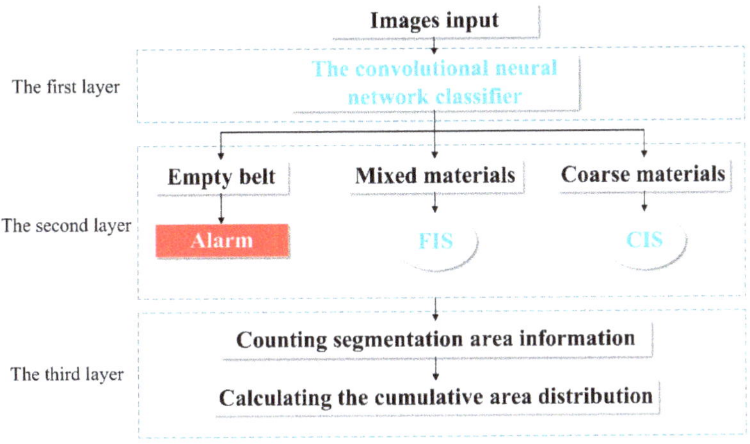

Figure 1. The technical roadmap of the method.

2.1. The Convolutional Neural Network Classifier

2.1.1. The Dataset Preparation

The size of the dataset is considered crucial for evaluating the performance of the trained model. An insufficient dataset causes a low recognition accuracy of the trained model. The dataset used in this study consisted of 2880 images collected from the process control system (PCS) system of a mineral-processing plant in the Yunnan Province, China. The images were taken by eight cameras installed on eight feeding belts with a collection rate of eight photos per minute for each camera. AXIS P3227-LVE cameras from AXIS are used, as well as LED PCS6-LED80 W lamps from Woodgrove. Two belt ore images are taken by each camera continuously every 15 s. The PCS system stores the latest 100 pictures from each camera; therefore, the images in the PCS system are completely updated every 12.5 min. We wrote a Python script to transfer the pictures from the storage folders to specified folders, and the transfer was executed every 13 min, lasting for a week. In the image transfer stage, the goal is to obtain sufficient belt ore images. It is efficient and economical to directly transfer images saved by the PCS system. At the same time, only two feeding belts are in running; therefore, there are a lot of duplicate images in the specified folders. All of the images have a size of 2304 × 1728 px (JPEG file). We planned to select about 3000 sharp, non-repetitive, and representative belt ore images as the original dataset. The 982 empty belt images were picked out including empty belts with water stains, empty belts with small particles, images taken by telephoto lens, and images taken by short focal length lens. We divided the belt ore images with fine content less than 30% into coarse material images. The features of coarse material images are obvious and similar; therefore, the number of coarse material images can be reduced appropriately. The 841 coarse material images were selected. The 1057 mixed material images were selected according to the proportion of fine material. The mixed material images were selected consisting of 100%, 90%, 70%, and 50% fine material. All images were taken from an industrial site and were not created; therefore, the proportion of fine material was an estimation and not an exact value. The 2880 high-quality, representative images were selected from the saved images as the original dataset. After physical verification, 522 px in each image was found to be equal to 30 cm. Several examples of the images are shown in Figure 2.

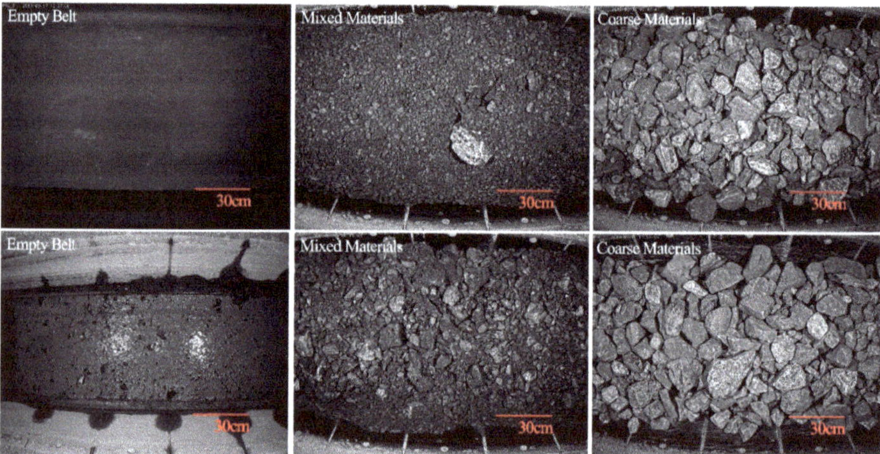

Figure 2. Samples of the original dataset: empty belt, mixed materials, and coarse materials.

2.1.2. Model Training

The design of our network is shown in Figure 3. The network consists of two convolution layers, two maxpool layers, two fully connected layers, and three ReLU activation functions. The DELL R730 server was used to train the model. The Windows Server 2012 was used as the operating system. The Intel E5–2609V4 was used as the CPU with a RAM of 32 GB, and the Nvidia P2000 (5 GB) was used as the GPU.

The original dataset was used as the raw data. The input images consisted of three channels, which were resized to 500 × 500 × 3. All input images were required to undergo a two-step pretreatment process. In the first step, the value range of pixels was changed from 0–255 to 0–1. In the second step, the image was normalized by using the empirical mean vector and the empirical std vector. The normalization is described as follows:

$$mean = [0.485, 0.456, 0.406] \tag{1}$$

$$std = [0.229, 0.224, 0.225] \tag{2}$$

$$result = (image - mean)/std \tag{3}$$

The original dataset was completely shuffled: 20% of the images assigned to the test set, 20% of the images assigned to the validation set, and 60% of the images used as the training set. The processed images were input into the neural network for model training. The epoch value was set to 10, the batch size was set to 32, the learning rate was set to 0.001, and the cross-entropy was used to evaluate the training loss. The prediction result was compared with the true label in order to calculate the training accuracy and the validation accuracy of every epoch. The training accuracy and validation accuracy were used to update the weights in the model [29]. The training process is shown in Figure 4. The training accuracy was 99.48%, the validation accuracy was 100%, and the training cross-entropy equaled 0.0146 when the epoch equaled 10. The model was tested with the test set, and the prediction accuracy was 100%.

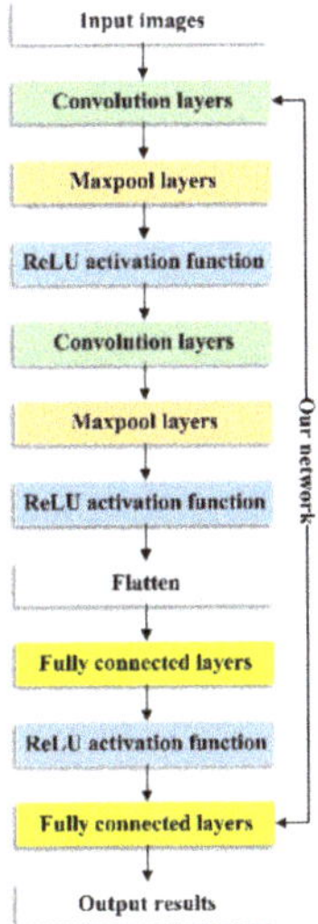

Figure 3. The architecture of our network.

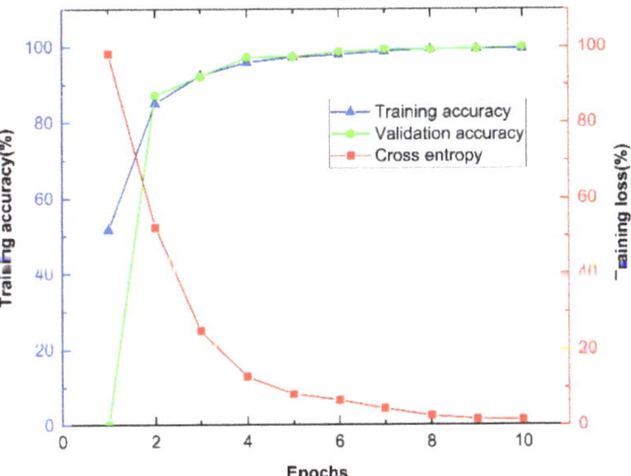

Figure 4. Training process using our network.

2.2. CIS and FIS Algorithms

2.2.1. CIS Algorithm

The CIS algorithm processes images classified by the classifier as coarse materials. The CIS algorithm is based on the Python OpenCV library (version 4.1.1). The uneven color distribution on the surface of the coarse ores and the coarse ores covered by fine particles led to the region of coarse ores in the image being divided into many small regions. Therefore, the CIS algorithm should be able to remove features that are similar to the ore edge on the surface of the coarse ore. The CIS algorithm is described in Figure 5.

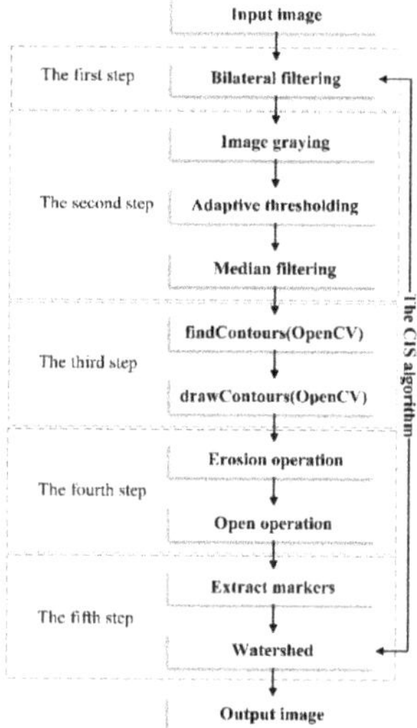

Figure 5. Block diagram of the coarse image segmentation (CIS) algorithm.

The CIS algorithm follows five steps. In the first step, the input image is processed using the bilateralFilter function. The setting value of the diameter of each pixel neighborhood equals 25, the sigma parameter of the color space equals 100, and the sigma parameter of the coordinate space equals 25. The purpose of bilateral filtering is to blur the surface of coarse materials while preserving the edge. The parameters of the bilateralFilter function were chosen according to experience and practice; for example, 25, 50, 75, 100. After bilateral filtering, the noise, details, and small color blocks on the surface of the ore are blurred, and the edge of the ore is preserved [30,31]. In the second step, the operations include graying [32], adaptive thresholding [33], and median filtering [34,35]. The setting kernel of the median filter equals 3. The kernel of the median filter should not be too high, to prevent edge interruption. After several tries, the results processed using the kernel with a value of 3 was more suitable for subsequent processing than 5 or 7. After the second step, the color image is binarized and denoised. In the third step, the image is processed using the findContours and drawContours functions. After these operations, the interconnected areas are closed. In the fourth step, the image is processed using the erode and morphologyEx functions [36–38]. The setting kernel is a 3 × 3 morph ellipse. The iterations parameter of the erode function is 10 and the iterations parameter of the morphologyEx function is 4. The target of erosion and open operation is to extract markers. The iterations parameters are sensitive, and are chosen by experience and trials. Finally, the watershed transformation is performed based on markers, and the watershed lines are drawn [10,39,40]. For instance, Figure 6 shows how the CIS algorithm processes a coarse material image.

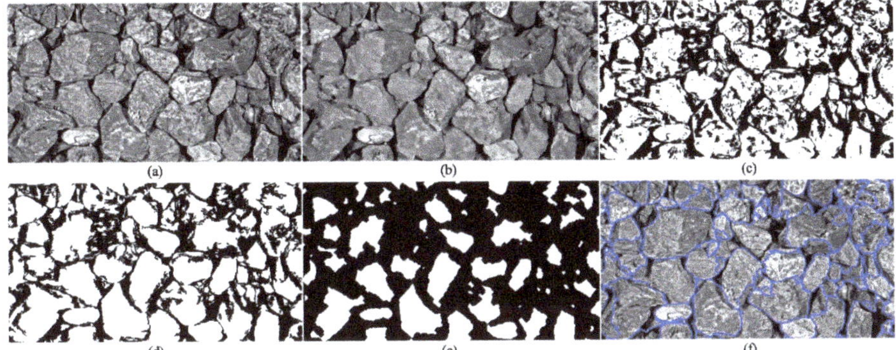

Figure 6. Original (**a**), after the first step (**b**), after the second step (**c**), after the third step (**d**), after the fourth step (**e**), and after the fifth step (**f**).

2.2.2. FIS Algorithm

The FIS algorithm processes the images classified by the classifier as mixed materials. The FIS algorithm is based on the Python OpenCV library (version 4.1.1). Both fine and coarse-fine materials are mixed materials. The edge of fine materials is weak; therefore, excessive blurring causes the under-segmentation of fine material images, and insufficient blurring causes the over-segmentation of coarse material images. The FIS algorithm must remove the edge-like features on the surface of coarse ores as much as possible while retaining the outlines of granular particles and fine material. The FIS algorithm is described in Figure 7.

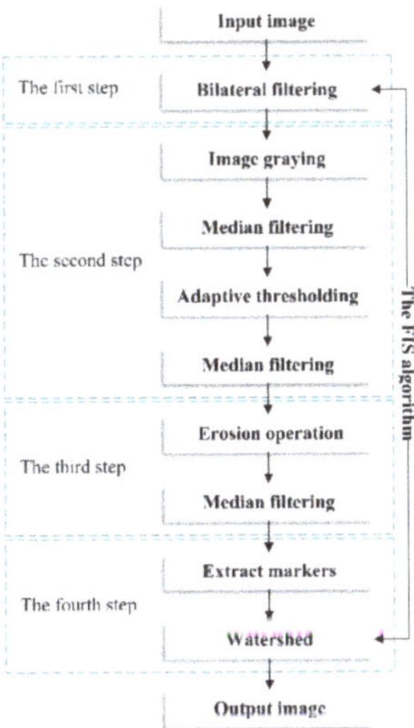

Figure 7. Block diagram of the fine image segmentation (FIS) algorithm.

The FIS algorithm follows four steps. In the first step, the input image is processed using the bilateralFilter function. Due to the need to deal with the coarse ores which are contained in the mixed materials, it is necessary to use bilateral filtering. Parameter settings are found to be the same as the CIS algorithm [30,31]. In the second step, the operations include graying [32], adaptive thresholding [33], and two median filterings [34,35]. The setting kernels of the median filters equal 3. After the second step, the color image is binarized and denoised. In the third step, the image is processed using the erode function and median filtering. The setting kernel of the median filter equals 9 and the kernel of the erosion operation is a 5 × 5 morph ellipse [38]. Due to the edge of fine materials being weak, we should not adopt strong morphological operations. Therefore, the iterations were set to one. To ensure complete separation among markers, the kernel of the median filter in the third step was set to 9. Finally, the watershed transformation was performed based on markers and the watershed lines were drawn [10,40]. A mixed material image was processed using the FIS algorithm as shown in Figure 8.

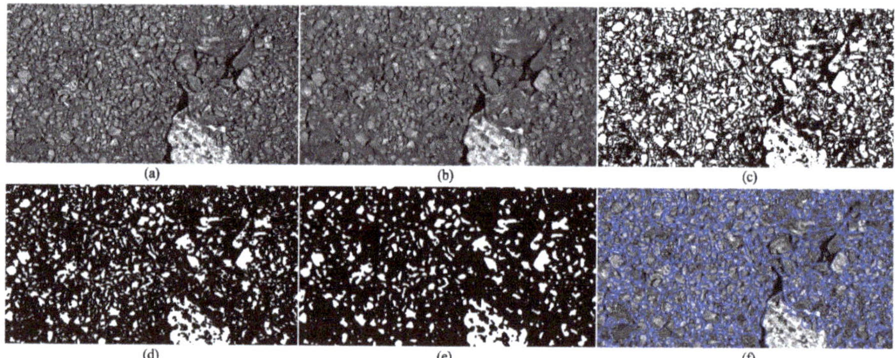

Figure 8. Original (**a**), after the first step (**b**), after the second step (**c**), after the erosion operation (**d**), after the third step (**e**), and after the fourth step (**f**).

3. Experimental Details

3.1. Image Acquisition and Classification

The experimental dataset covers 12 representative images which consist of four empty belt images labelled as one, four mixed material images labelled as two, and four coarse material images labelled as zero. All images were taken from the PCS system at different times. We selected four empty belt images including an empty belt with water stains, an empty belt with small particles, an image taken by a telephoto lens, and an image taken by a short focal length lens. The features of coarse material images are obvious and similar. The four coarse material images were selected randomly. The four mixed material images were selected consisting of 100%, 90%, 70%, and 50% fine material. The raw images were divided into the following four groups: one empty belt image, one mixed material image, and one coarse material image in each group. In order to compare the segmentation results of the different algorithms, the designated region of raw images was processed by using different algorithms. The size of the designated region was 1202 × 631 (JPEG file). Manual segmentation was used to obtain accurate segmentation images. Images from group one (as shown in Figure 9a) were processed using PCS, CIS, and FIS, respectively, for evaluating each algorithm. Other images were processed and evaluated by using our method.

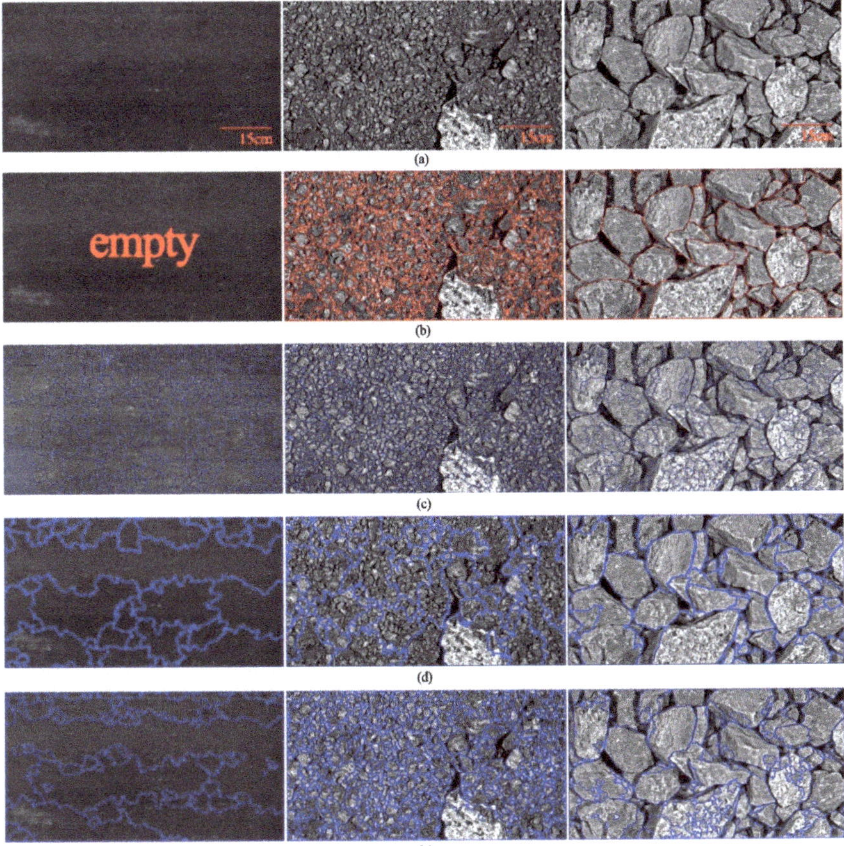

Figure 9. The segmentation results processed by different algorithms: the original images (**a**), manual segmentation (**b**), process control system (PCS) (**c**), CIS (**d**), and FIS (**e**).

3.2. Estimation of Segmentation Accuracy

The area (px) of the regions surrounded by contours in the segmented images were counted, and the cumulative area distribution was calculated. Suppose $x_1, x_2, \ldots, x_n$ are areas of the regions enclosed by the particle segmentation contours in a processed image. Among $x_1, x_2, \ldots, x_n$, areas of the regions that are smaller than the specified values are $y_1, y_2, \ldots, y_m$. The cumulative area distribution (C) is computed using the following equation:

$$C = \frac{\sum_{i=1}^{m} y_i}{\sum_{j=1}^{n} x_j} \quad (m \leq n) \tag{4}$$

We use the cumulative area distribution to evaluate the segmentation accuracy of algorithms. The specified value used to calculate the cumulative area distribution is called the area filter (px). If the number of pixels in the region surrounded by the contour is less than the area filter, the region can pass the area filter; otherwise, the region cannot pass. The contourArea function from the Python OpenCV library (version 4.1.1) was adopted to get the number of pixels in the region surrounded by the contour.

The cumulative area distributions of the segmentation results of the empty belt image, mixed material image, and coarse material image are quite different; therefore, different area filters were used for calculating the cumulative area distribution of the segmented images. The empty belt images were evaluated using 2000, 4000, 8000, 10,000, 20,000, 40,000, 80,000, and 100,000 px area filters. The mixed material images were evaluated using 2000, 4000, 6000, 8000, 10,000, 20,000, 30,000, and 40,000 px area filters. The coarse material images were evaluated using 5000, 10,000, 15,000, 20,000, 25,000, 30,000, 35,000, and 40,000 px area filters. The cumulative area distribution and the number of segmentation contours of the image segmented by the PCS system, CIS, and FIS, respectively, were compared with the segmentation result of the manual segmentation image. The segmentation result of the algorithm, which was close to the manual segmentation result, was evaluated as accurate.

4. Experimental Results and Discussions

4.1. The Segmentation Result Analysis of Different Algorithms

Images in Figure 9a were segmented using manual segmentation, PCS, CIS, and FIS, respectively, providing results as shown in Figure 9b–e. By counting the number of segmentation contours in the segmented images, the results are shown in Table 1. By calculating the cumulative area distribution of the segmented images, the results are shown in Figure 10. Table 1 shows that the segmentation contour counts of the empty belt image by the PCS system, CIS algorithm and FIS algorithm, respectively, are 371, 51, 92. Figure 10a and Table 1 demonstrate that no algorithm can accurately segment the empty belt images. Due to many disturbing factors such as small particles, dirt, and water on the surface of the empty belts, the segmentation algorithms always segments the empty belt images. We only require to recognize and need not segment the empty belt images. Table 2 shows our method alarms all three empty belt images from groups 2 to 4. Table 2 indicates that using the training model based on the convolutional neural network to process the empty belt images shows better performance and accuracy. From Figure 10c and Table 1, it is observed that the PCS system shows serious over-segmentation in processing the coarse material images, and the segmentation accuracy of the FIS algorithm is found to be between that of the CIS algorithm and the PCS system. The outlines of massive rocks are obvious as compared to those of granular particles and fine material. Many disturbing factors, such as edges and corners, uneven color distribution, shadow, and so on, are found on the surface of massive rocks. An algorithm that can accurately segment the coarse material images must overcome the above interferences, however, the blur and denoise operations used cannot weaken the outlines of coarse ores too much. Figure 9 shows that the CIS algorithm has obvious advantages for segmenting the coarse material images. The blur, denoise, and smooth operations of the CIS algorithm weaken the outlines of ores; therefore, there is a serious under-segmentation in processing mixed material images, especially the fine material images (see Figures 9 and 10b). We classified the images in the original dataset, which are neither coarse material images nor empty belt images, as mixed material images.

Table 1. The segmentation contour count of various algorithms.

Classification	Segmentation Contour Count			
	Manual	PCS	CIS	FIS
Empty belt	0	371	51	92
Mixed materials	885	942	123	955
Coarse materials	101	408	91	280

Table 2. The results of empty belt images from groups 2 to 4, as processed by our method.

Groups	Status
2	Alarm
3	Alarm
4	Alarm

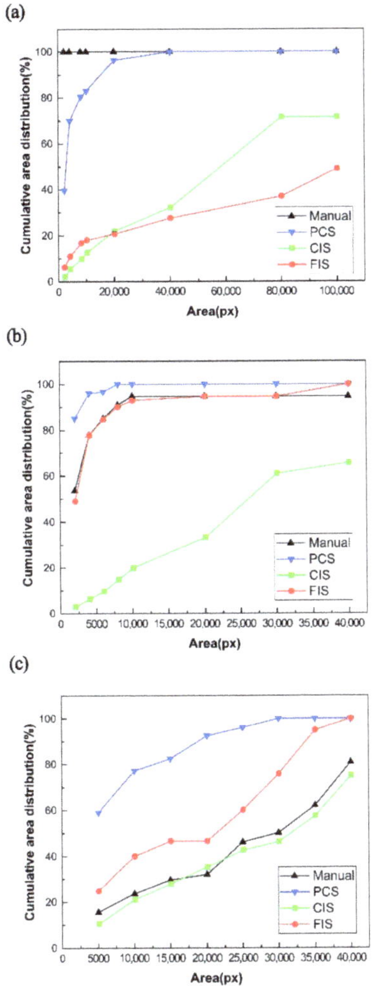

Figure 10. The results of raw images by segmenting group 1 using manual segmentation, PCS, CIS, and FIS, respectively: empty belt (**a**), mixed materials (**b**), and coarse materials (**c**).

As the number of images in the original dataset is very high, it is not feasible to count the particle size distribution of materials on conveyor belts. The method of distinguishing empty, mixed materials, and coarse materials is based on artificial classification. An image that is full of massive ores is classified as a coarse material image. The accurate segmentation of mixed material images is most difficult to achieve. The massive ores show a greater impact on the grinding process. The results calculated by the under-segmentation of fine material images are found to be much coarser than the actual results. Fewer morphological operations enable the FIS algorithm to retain the outlines of granular particles and fine materials. More denoise operations enable the FIS algorithm to remove the interferences on the surface of the massive ores as much as possible. Figure 10b shows that the FIS algorithm shows accurate segmentation in processing mixed material images, especcially the fine material images. As the FIS algorithm cannot remove the large non-edge features on the surface of massive rocks such as edges, corners, and color blocks, the FIS algorithm does not show as accurate results as the CIS algorithm in processing the coarse material images (see Figure 10c and Table 1). It is more accurate and reasonable to classify images first and then use different algorithms for processing rather than to process all images with the same algorithm.

4.2. The Processing Result Analysis of Our Method

Images from groups 2 to 4 were processed by using both our method and manual segmentation. Three mixed material images include one fine material image from group 2 and two mixed material images from groups 3 and 4. The results of the three processed empty belt images are shown in Table 2. The results of the three processed mixed material images are shown in Table 3. The results of the three processed coarse material images are shown in Table 4. The cumulative area distribution of one fine material image from group 2, one mixed material image from group 3, and one coarse material image from group 2, which were processed by our method, are shown in Figure 11. The column headers in Tables 3 and 4 (5000, 10,000, etc.) indicate area filters. Table 2 shows that three empty belt images used for testing were alarmed. Our method is found to be accurate and reliable for empty belt identification. From Table 3, it is observed that the maximum error of the cumulative area distribution calculated by using different area filters is 2.71% for fine material, 4.65% for mixed materials from group 3, and 5.02% for mixed materials from group 4. From Table 3, the average error of the cumulative area distribution calculated by using different area filters is 1.07% for fine material, 2.27% for mixed materials from group 3, and 2.89% for mixed materials from group 4. From Table 4, it is observed that the maximum error of the cumulative area distribution calculated by using different area filters is 3.51%, 5.61%, 3.83%, respectively, for coarse materials from groups 2 to 4. From Table 4, the average error of the cumulative area distribution calculated by using different area filters is 1.30%, 3.30%, 2.59%, respectively, for coarse materials from groups 2 to 4. Tables 3 and 4 indicate that our method can segment both mixed material images and coarse material images with high precision. Our method is considered useful for identifying and segmenting the images taken on industrial conveyor belts.

Table 3. The results of mixed material images from groups 2 to 4, processed by our method.

Groups	Methods	Cumulative Area Distribution (%)							
		2000	4000	6000	8000	10,000	20,000	30,000	40,000
2 (F)	Our method	55.59	85.64	96.82	100	100	100	100	100
	Manual	58.06	87.6	94.11	98.61	100	100	100	100
3 (M)	Our method	16.44	36.67	50.89	62.29	69.54	89.07	92.08	92.08
	Manual	18	38.09	55.54	66.82	73.67	89.49	92.8	92.8
4 (M)	Our method	26.4	50.64	67.1	77.11	90.78	95.17	100	100
	Manual	28.09	55.66	71.1	80.66	94.77	100	100	100

F: fine material; M: mixed materials.

Table 4. The results of coarse material images from groups 2 to 4, processed by our method.

Groups	Methods	Cumulative Area Distribution (%)							
		5000	10,000	15,000	20,000	25,000	30,000	35,000	40,000
2	Our method	8.86	26	41.58	56.73	63.02	72.61	79.91	85.05
	Manual	10.43	29.51	41.94	57.48	64.07	74.07	80.65	85.99
3	Our method	6	15.38	25.81	40.83	57.06	64.08	79.86	84.57
	Manual	7.34	18.55	31.42	43.26	61.66	67.26	85.26	85.26
4	Our method	8.49	22.74	33.68	43.76	51.15	66.5	71	74.34
	Manual	6.92	25.5	37.02	46.46	53.02	63.73	72.86	78.17

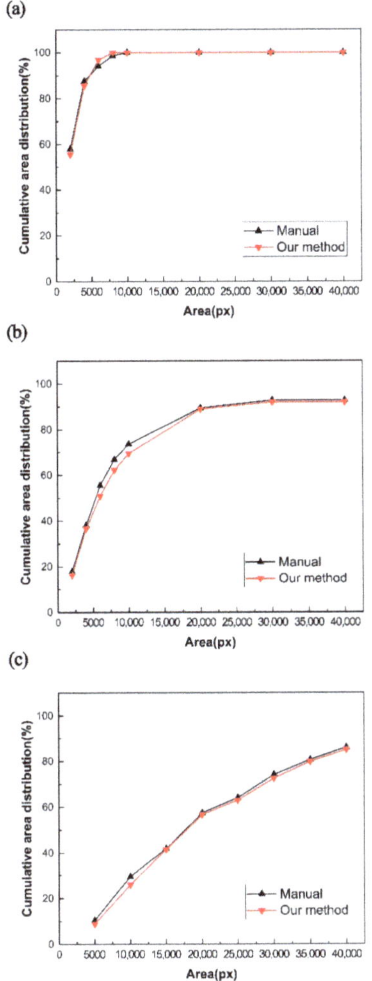

Figure 11. The results of the specified images processed by our method: fine material (**a**), mixed materials (**b**), and coarse materials (**c**).

5. Conclusions

The objective of this study is to develop a method that can accurately segment belt ore images. The accurate image segmentation method is considered important for estimating the size distribution of mineral materials on industrial conveyor belts. For that purpose, 2880 images collected from a process control system on the industrial site were processed as the original dataset. Deep learning and image processing techniques were integrated with the image segmentation method. From the perspective of an application, the accurate recognition of the empty belts is necessary to realize the automatic switch of conveyor belts. The new method can identify the empty belt images with high precision. Moreover, it is difficult for an image segmentation algorithm to achieve accurate segmentation of both coarse materials and mixed materials. This study adopted the convolutional neural network model to solve the automatic classification of belt ore images, and then used the CIS algorithm and the FIS algorithm to accurately segment coarse material images and mixed material images, respectively. The new method makes it feasible and efficient to accurately process various belt ore images. Notably, it can be used as a brand new algorithm for belt ore image processing. The main novelties are as follows:

- This study used a convolutional neural network to identify empty belts.
- The new method adopted the strategy which is to classify the belt ore images first and then use different algorithms for processing different kinds of images.

Author Contributions: Conceptualization, X.M. (Xiqi Ma) and L.O.; methodology, X.M. (Xiqi Ma); software, X.M. (Xiqi Ma) and P.Z.; validation, X.M. (Xiqi Ma) and X.M. (Xiaofei Man); formal analysis, X.M. (Xiqi Ma) and P.Z.; investigation, X.M. (Xiqi Ma); resources, L.O.; data curation, X.M. (Xiqi Ma); writing—original draft preparation, X.M. (Xiqi Ma), X.M. (Xiaofei Man) and P.Z.; writing—review and editing, X.M. (Xiqi Ma), P.Z. and L.O.; supervision, L.O. All authors have read and agreed to the published version of the manuscript.

Funding: This work was financially supported by the National Natural Science Foundation of China (No. 51674291).

Acknowledgments: The authors also thank the support of the Key Laboratory of Hunan Province for Clean and Efficient Utilization of Strategic Calcium-containing Mineral Resources (No. 2018TP1002). Special thanks to Wencai Zhang for his suggestions on paper writing.

Conflicts of Interest: The authors declare no conflict of interest.

References

1. Tessier, J.; Duchesne, C.; Bartolacci, G. A machine vision approach to on-line estimation of run-of-mine ore composition on conveyor belts. *Miner. Eng.* **2007**, *20*, 1129–1144. [CrossRef]
2. Lange, T.B. Real-time measurement of the size distribution of rocks on a conveyor belt. *IFAC Proc. Vol.* **1988**, *21*, 25–34. [CrossRef]
3. Ko, Y.D.; Shang, H. A neural network-based soft sensor for particle size distribution using image analysis. *Powder Technol.* **2011**, *212*, 359–366. [CrossRef]
4. Hamzeloo, E.; Massinaei, M.; Mehrshad, N. Estimation of particle size distribution on an industrial conveyor belt using image analysis and neural networks. *Powder Technol.* **2014**, *261*, 185–190. [CrossRef]
5. Lin, C.L.; Miller, J.D. Development of a pc, image-based, on-line particle-size analyzer. *Min. Metall. Explor.* **1993**, *10*, 29–35. [CrossRef]
6. Liao, C.W.; Tarng, Y.S. On-line automatic optical inspection system for coarse particle size distribution. *Powder Technol.* **2009**, *189*, 508–513. [CrossRef]
7. Yen, Y.K.; Lin, C.L.; Miller, J.D. Particle overlap and segregation problems in on-line coarse particle size measurement. *Powder Technol.* **1998**, *98*, 1–12. [CrossRef]
8. Singh, V.; Rao, S.M. Application of image processing and radial basis neural network techniques for ore sorting and ore classification. *Miner. Eng.* **2005**, *18*, 1412–1420. [CrossRef]
9. Al-Thyabat, S.; Miles, N.J.; Koh, T.S. Estimation of the size distribution of particles moving on a conveyor belt. *Minerals Eng.* **2007**, *20*, 72–83. [CrossRef]
10. Ilya, L.; Hong, Z. Classification-driven watershed segmentation. *IEEE Trans. Image Process.* **2007**, *16*, 1437–1445.
11. Outal, S.; Schleifer, J.; Pirard, E. In Evaluating a calibration method for the estimation of fragmented rock 3d-size-distribution out of 2d images. In Proceedings of the FRAGBLAST 9—9th International Symposium on Rock Fragmentation by Blasting, Granada, Spain, 13–17 September 2009; pp. 221–228.
12. Andersson, T.; Thurley, M.J. Minimizing profile error when estimating the sieve-size distribution of iron ore pellets using ordinal logistic regression. *Powder Technol.* **2011**, *206*, 218–226. [CrossRef]
13. Igathinathane, C.; Pordesimo, L.O.; Columbus, E.P.; Batchelor, W.D.; Methuku, S.R. Shape identification and particles size distribution from basic shape parameters using imagej. *Comput. Electron. Agric.* **2008**, *63*, 168–182. [CrossRef]
14. Gawenda, T.; Krawczykowski, D.; Krawczykowska, A.; Saramak, A.; Nad, A. Application of dynamic analysis methods into assessment of geometric properties of chalcedonite aggregates obtained by means of gravitational upgrading operations. *Minerals* **2020**, *10*, 180. [CrossRef]
15. Krawczykowski, D. Application of a vision systems for assessment of particle size and shape for mineral crushing products. *IOP Conf. Ser. Mater. Sci. Eng.* **2018**, *427*, 1–5. [CrossRef]
16. Fannin, R.J.; Shuttle, D.A.; Rousé, P.C. Influence of roundness on the void ratio and strength of uniform sand. *Géotechnique* **2008**, *58*, 227–231.

17. Split Engineering Products. Split-Online Software and Systems. 2020. Available online: https://www.spliteng.com/products/split-online-systems/ (accessed on 12 January 2020).
18. Outotec Products and Services. Outotec®Act Grinding Optimization System. 2020. Available online: http://www.outotec.cn/products-and-services/technologies/grinding/act-grinding-optimization-system/ (accessed on 3 February 2020).
19. WipWare Products and Services. Wipware Conveyor Analysis System. 2020. Available online: http://wipware.com/products/momentum/ (accessed on 7 January 2020).
20. Zhang, Z.; Yang, J.; Ding, L.; Zhao, Y. Estimation of coal particle size distribution by image segmentation. *Int. J. Mining Sci. Technol.* **2012**, *22*, 739–744.
21. Chauhan, R.; Ghanshala, K.K.; Joshi, R.C. Convolutional neural network (cnn) for image detection and recognition. In Proceedings of the 2018 First International Conference on Secure Cyber Computing and Communication (ICSCCC), Jalandhar, India, 15–17 December 2018.
22. Sultana, F.; Sufian, A.; Dutta, P. Advancements in image classification using convolutional neural network. In Proceedings of the 2018 Fourth International Conference on Research in Computational Intelligence and Communication Networks (ICRCICN), Kolkata, India, 22–23 November 2018; IEEE: Piscataway, NJ, USA; pp. 122–129.
23. Krizhevsky, A.; Sutskever, I.; Hinton, G.E. Imagenet classification with deep convolutional neural networks. *Adv. Neural Inf. Process. Syst.* **2012**, *25*. [CrossRef]
24. He, K.; Zhang, X.; Ren, S.; Sun, J. Deep residual learning for image recognition. In Proceedings of the IEEE Conference on Computer Vision and Pattern Recognition, Las Vegas, NV, USA, 27–30 June 2016; pp. 770–778.
25. Szegedy, C.; Liu, W.; Jia, Y.; Sermanet, P.; Reed, S.; Anguelov, D.; Erhan, D.; Vanhoucke, V.; Rabinovich, A. Going deeper with convolutions. In Proceedings of the IEEE Conference on Computer Vision and Pattern Recognition, Columbus, OH, USA, 23–28 June 2014.
26. Simonyan, K.; Zisserman, A. Very deep convolutional networks for large-scale image recognition. *arXiv* **2014**, arXiv:1409.1556.
27. Ye, L.; Chao, G.; Cheng, G. Rock classification based on images color spaces and artificial neural network. In Proceedings of the Fifth International Conference on Intelligent Systems Design & Engineering Applications, Hunan, China, 15–16 June 2014.
28. Cheng, G.; Guo, W. Rock images classification by using deep convolution neural network. *J. Physics Conf. Ser.* **2017**, *887*, 012089. [CrossRef]
29. Liu, C.; Li, M.; Zhang, Y.; Han, S.; Zhu, Y. An enhanced rock mineral recognition method integrating a deep learning model and clustering algorithm. *Minerals* **2019**, *9*, 516. [CrossRef]
30. Guarnieri, G.; Marsi, S.; Ramponi, G. Fast bilateral filter for edge-preserving smoothing. *Electron. Lett.* **2006**, *42*, 396–397. [CrossRef]
31. Tomasi, C.; Manduchi, R. Bilateral filtering for gray and color images. In Proceedings of the International Conference on Computer Vision, Copenhagen, Denmark, 28–31 May 2002.
32. Zhang, X.; Wang, X. Novel survey on the color-image graying algorithm. In Proceedings of the IEEE International Conference on Computer & Information Technology, Helsinki, Finland, 21–23 August 2017.
33. Bradley, D.; Roth, G. Adaptive thresholding using the integral image. *J. Graph. GPU Game Tools* **2007**, *12*, 13–21. [CrossRef]
34. Tang, J.; Wang, Y.; Cao, W.; Yang, J. Improved adaptive median filtering for structured light image denoising. In Proceedings of the International Conference on Information, Communication and Networks, Macau, China, 24–26 April 2019.
35. Ataman, E.; Aatre, V.K.; Wong, K.M. A fast method for real-time median filtering. *IEEE Trans. Acoust. Speech Signal. Process.* **1980**, *28*, 415–421. [CrossRef]
36. Lien, B.K. Efficient implementation of binary morphological image processing. *Opt. Eng.* **1994**, *33*, 3733–3738. [CrossRef]
37. Sha, H.; Wah, C.C. Morphological image processing and its parallel implementation. In Proceedings of the 3rd International Conference on Signal Processing, Beijing, China, 18 October 1996.
38. Gonzalez, R.C.; Woods, R.E.; Eddins, S.L. *Digital Image Processing Using Matlab*; Pearson Education India: Noida, India, 2004.

39. Amankwah, A.; Aldrich, C. Rock image segmentation using watershed with shape markers. In Proceedings of the IEEE 39th Applied Imagery Pattern Recognition Workshop (AIPR), Washington, DC, USA, 13–15 October 2010.

40. Smet, P.D. Implementation and analysis of an optimized rainfalling watershed algorithm. In Proceedings of the International Society for Optics and Photonics, San Jose, CA, USA, 19 April 2000.

Publisher's Note: MDPI stays neutral with regard to jurisdictional claims in published maps and institutional affiliations.

 *minerals*

Article

Assessment of Sortability Using a Dual-Energy X-ray Transmission System for Studied Sulphide Ore

YiRan Zhang, Nawoong Yoon and Maria E. Holuszko *

The Norman B. Keevil Institute of Mining Engineering, University of British Columbia, Vancouver, BC V6T 1Z4, Canada; yiran.zhang@alumni.ubc.ca (Y.Z.); nawoong.yoon@alumni.ubc.ca (N.Y.)
* Correspondence: maria.holuszko@ubc.ca

Abstract: In hard rock mining, sensor-based sorting can be applied as a pre-concentration method before the material enters the mill. X-ray transmission sensors have been explored in mining since 1972. Sorting ore of acceptable grade and waste material before processing at the mill can reduce the amount of tailings per unit of valuable metal in the mining operation and have many economic benefits. Ore samples used in this paper are from a polymetallic carbonate replacement deposit (gold-silver-lead-zinc sulphide) in Southeast Europe. This paper focuses on how the Dual-Energy X-ray Transmission (DE-XRT) data is generated and used for ore characterization and sortability for this sulphide ore. The method used in the DE-XRT analysis in this project is based on the dual-material decomposition method, which is used in the medical industry for radiology. This technique can distinguish sulphides from non-sulphides. However, the correlation developed between the DE-XRT response and the metal content is lacking. As a result, the DE-XRT response can only classify the material effectively but cannot reliably predict the metal content.

Keywords: Dual-Energy X-ray Transmission (DE-XRT); sulphide; polymetallic

Citation: Zhang, Y.; Yoon, N.; Holuszko, M.E. Assessment of Sortability Using a Dual-Energy X-ray Transmission System for Studied Sulphide Ore. *Minerals* **2021**, *11*, 490. https://doi.org/10.3390/min11050490

Academic Editors: Andreas Delimitis and Daniel Saramak

Received: 8 March 2021
Accepted: 1 May 2021
Published: 4 May 2021

Publisher's Note: MDPI stays neutral with regard to jurisdictional claims in published maps and institutional affiliations.

1. Introduction

Ore sorting is a sensor-based sorting and pre-concentration technology that is implemented before the processing stage and used to reduce the amount of low-grade ore and waste reporting to the mill feed. X-ray transmission sensors have been explored in mining since 1972, as described by Jenkinson et al. [1] This technology is widely used at airports for baggage inspection and the basic principles have been adopted as a sorting technique [2] This technology can also be utilized to recover ore from previously uneconomic waste and reduce the mill energy and reagent consumption by reducing the mass of ore being processed by the mill [3]. In greenfield projects, sensor-based sorting can further provide value by lowering processing capital costs [4]. The decline of available high-grade orebodies and decreasing head grades has further increased the attractiveness of pre-concentration, and, subsequently, ore sorting [5].

A critical parameter in obtaining the DE-XRT result is developing the H-L (High and Low energy) curves that determine the relative density of a given pixel based on its X-ray attenuation. In this paper, a dual-material decomposition method will be used to determine the sulphide material from the waste material. This study aims to assess the effectiveness of DE-XRT to sort the sulphides from the non-sulphides. Since the ore is very heterogeneous, different rock types were identified to account for the mineral composition of various pieces of rock. Hence, the test work was designed to assess each rock representing different mineralogy or rock type. Density measurements using DE-XRT were used to determine average atomic density by measuring the X-ray attenuation for each specimen used in the study, respectively. Both silver and gold assays are determined by the fire assay method, and the rest of the elements are determined by the Induction Coupled Plasma Method (ICP).

In this study, the equivalent gold grade is used to assess the value of the sample since it is a polymetallic deposit. The equivalent gold grade is calculated by summarizing the dollar value of gold, silver, zinc, and lead, divided by the price of gold. The commodity price used in this study to calculate equivalent gold grade is 1300 \$/oz of Au, 17.5 \$/oz of Ag, 5 \$/lb of Mo, 1.25 \$/lb of Zn and 1.05 \$/lb of Pb.

2. Materials and Methods

Three main parameters were recorded for each rock specimen in this experiment: rock classification, DE-XRT output, and assays. It is critical to ensure that the same sample is assessed for each parameter to obtain correct data for correlation analysis. Five hundred rock specimens were used in this study, and the particle size ranged between +50 mm to 100 mm. The initial sample consisted of run-of-mine (ROM) material, which included fines and coarse particles not suitable for particle sorting. The ROM material was screened to the desired particle size between 50 mm and 100 mm and washed where the five hundred representative samples were selected.

2.1. Material Classification

Individual rock specimen has been classified into two classes based on visual observations of mineralization. When classifying rock types into the two classes, two factors are considered for this ore deposit. Initially, visual parameters are considered where rock specimens are observed showing mineralization; those with silver or golden tints are considered sulphides, while the dull and whitish-looking rocks are considered non-sulphides (mostly carbonates in this case). Also, the density of the samples aids the classification, where the heavier sample of the same rock size is classified as a sulphide. Examples of sulphide and non-sulphide specimens are illustrated in Figure 1.

Figure 1. Non-sulphides (**left**) and sulphides (**right**).

2.2. Dual-Energy X-ray Transmission (DE-XRT)

The Dual-Energy X-ray Transmission (DE-XRT) technology measures the X-ray attenuation according to Beer's Law for monochromatic narrow X-ray beams [6], which is expressed in the equation below. The final μ coefficient will be a function of radiation energy in addition to the properties like material density and thickness [7].

$$H = e^{-\mu_H t} \tag{1}$$

$$L = e^{-\mu_L t} \tag{2}$$

μ_H : Mass attenuation coefficient for High Energy (cm^2/g)
μ_L : Mass attenuation coefficient for Low Energy (cm^2/g)
t : Mass thickness (g/cm^2)

The relationship between the high- and low-energy X-ray attenuation can be expressed by rearranging Equations (1) and (2), which is the basis of H-L curve generation:

$$H = L^{\frac{\mu_H}{\mu_L}} \tag{3}$$

In this experiment, COMEX's MSX-400-VL-XR system installed at the University of British Columbia (UBC) was used and illustrated in Figure 2. The equipment has an X-ray sensor with a 1.5 mm resolution and a belt width of 51 cm travelling at a speed of 0.5 m/s. In order to minimize the error from equipment malfunction such as miscalibrated detectors, the machine is maintained regularly and necessary repair would need to be completed [8].

Figure 2. COMEX's MSX-400-VL-XR System in UBC's Coal and Mineral Processing Laboratory.

Using the COMEX MSX-400-VL-XR system, H-L curves are generated. The low-energy X-ray response represents the x-axis, and the high-energy X-ray response represents the y-axis. To use the dual-material decomposition method, two distinct material types were chosen. Figure 3 illustrates a visual representation of the dual-material decomposition method where Material 1 and Material 2 are vectors and where material can be represented as a sum of Material 1 and Material 2.

$$\vec{M} = \vec{A} + \vec{B} \tag{4}$$

where $\vec{M}$ *is unknown material*
$\vec{A}$ *is Material Type 1*
$\vec{B}$ *is Material Type 2*

In this study, the two types of material chosen for validation were dictated by the amount of equivalent gold content in the specimen. The rock with an equivalent gold grade in the top fifth percentile (sulphide) and the bottom fifth percentile (non-sulphide) was chosen to generate the characteristic function in the H-L plot. The average grade of the entire sample was 13.47 g/t gold, 80.32 g/t silver, 2.58% lead, and 2.66% zinc. The unit of 1 g/t was equivalent to 1 ppm since there are one million grams in one metric ton. The rocks classified in the top fifth percentile of metal content can have gold grades as high as 60 g/t, silver grades of 200 g/t, lead grades of 7%, or zinc grades of 4%.

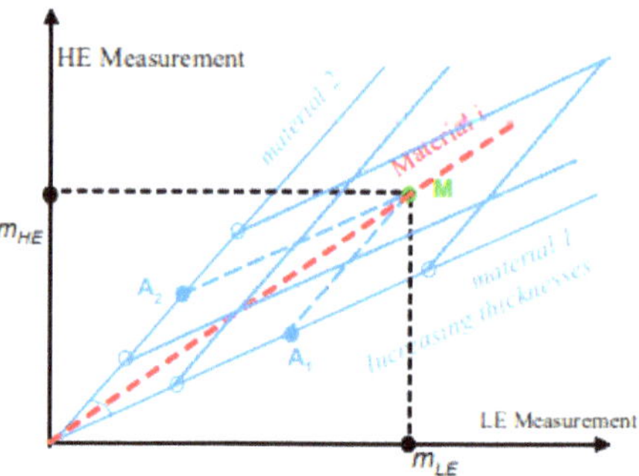

Figure 3. The material decomposition method using vector addition [9].

The relative density number for each pixel is determined by finding the magnitude of each vector from sulphides and non-sulphides. The relative density value, in this case, represents the percentage of being rock referred to as Material type 1, which is expressed in the equation below.

$$Relative\ Density = \frac{Magnitude\ of\ \vec{A}}{Magnitude\ of\ \vec{A} + Magnitude\ of\ \vec{B}} \tag{5}$$

$\vec{A}$ = *Material Type 1*
$\vec{B}$ = *Material Type 2*

2.3. Assays

Upon completion of the XRT tests, the rocks were then sent for assaying. The assay determined the amount of valuable metals present in every rock using 30 g of a sample for fire assay. Additionally, a 33-element Inductively Coupled Plasma Atomic Emission Spectroscopy (ICP-AES, MS Analytical, Lanley, B.C., Canada was conducted on the individual samples.

3. Results and Discussion

In this sample, the ore and waste have been determined by the amount of particular mineralization visible macroscopically (by the naked eye). Of the five hundred rock specimens, 218 (43.6%) samples were visually classified as non-sulphides, and the remaining 282 (46.4%) samples were sulphides.

Figure 4 illustrates the distribution of six different assays for all samples, with rock types differentiated by colour (orange sulphides; blue non-sulphides). The histograms show that sulphide material tends to have a much higher grade for gold, silver, zinc, and lead, while non-sulphide materials tend to have much higher calcium content.

As illustrated in Figure 5, there are two distinct sets of materials. Sulphides (orange dots) have high equivalent gold grades as well as high sulphur content, while the non-sulphides have low equivalent gold grades and low sulphur content. This infers that there are distinct characteristics between the sulphide material and non-sulphide material.

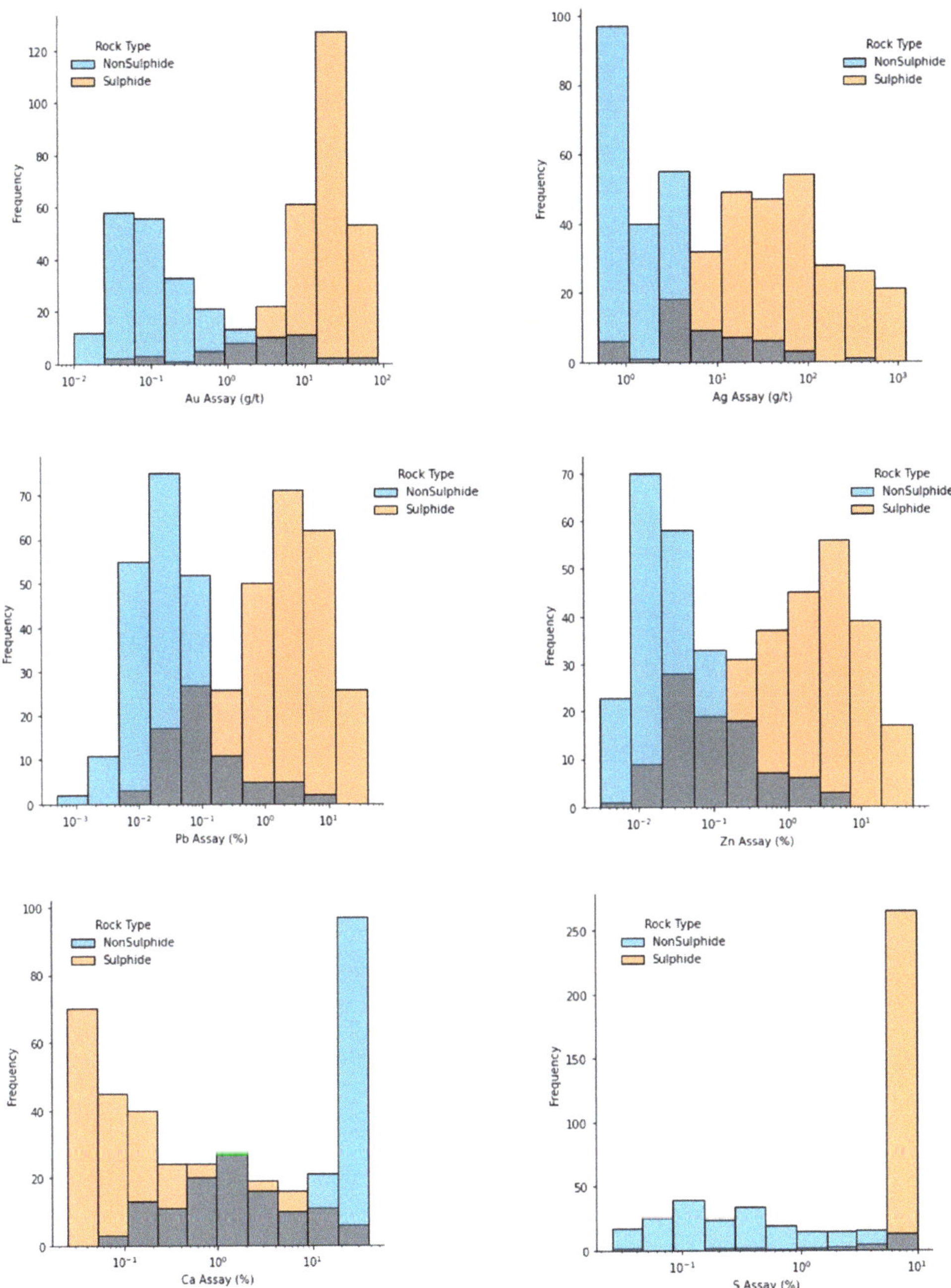

Figure 4. Histogram of six different assays based on rock type.

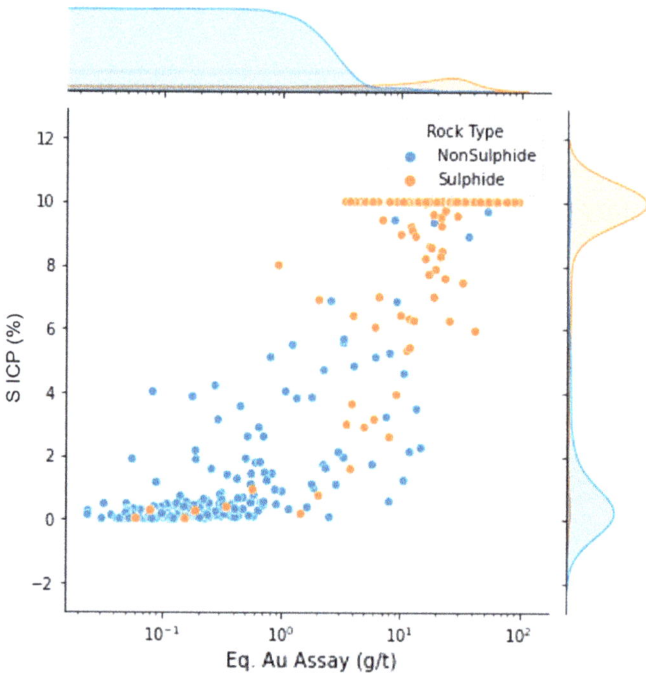

Figure 5. Scatter plot of equivalent gold assay versus ICP sulphur grade. Eq. Au—Equivalent Gold.

3.1. DE-XRT Image Processing

Based on the H-L curve generated using the top and bottom 5 % of the samples with the highest equivalent gold grade, each specimen produced an image with relative density values representing each pixel. Three samples selected in Table 1 illustrate the difference between high, medium and low sulphur content samples (indicating sulphide content). The pixel values range from 0 to 1, where 0 is represented as blue, and 1 is represented as yellow.

In Table 1, there are distinct characteristics between these three samples. The colours represent each pixel's relative density, ranging from 0 (dark blue) to 1 (yellow). The high relative density indicates that the sample is more likely to be a sulphide material. Valuable metals, iron, and sulphur are proportionally correlated, while the calcium content is inversely correlated. The high calcium content indicates that the waste material is composed of carbonates, while the sulphides have high iron, sulphur, and valuable metal content.

It is easy to distinguish the high sulphur samples from the low sulphur samples based on the relative XRT density distribution as illustrated in Figure 6. This suggests that taking an average value of the relative density of an entire sample can be effective in separating the sulphide from the non-sulphide rocks.

Table 1. Comparison of three different specimens with low, medium, and high sulphur content.

Sample Number	260	493	459
Images			
Au Assay (g/t)	0.34	6.93	41.19
Ag Assay (g/t)	0.50	40.00	347.00
Pb Assay (%)	0.12	1.32	10.01
Zn Assay (%)	0.22	3.68	6.05
Eq. Au Assay (g/t)	0.55	10.62	55.39
As ICP (%)	0.21	7.92	10.00
Ca ICP (%)	27.09	8.87	0.09
Fe ICP (%)	0.72	7.37	23.86
Mg ICP (%)	1.65	4.54	0.03
Mn ICP (%)	2.25	1.40	0.10
S ICP (%)	0.53	4.60	10.00
Average Relative Density	0.52	0.54	0.91

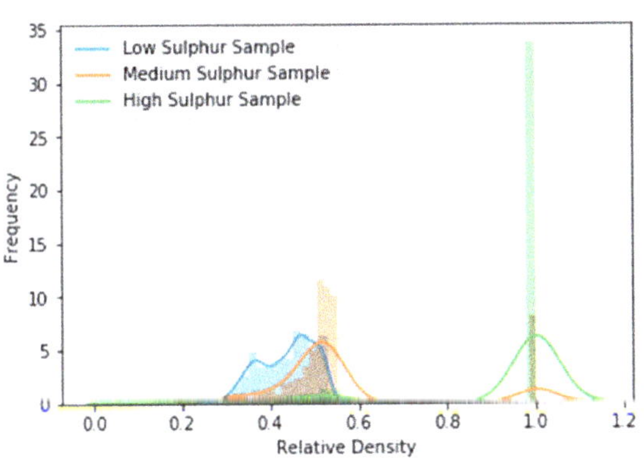

Figure 6. Histogram of relative XRT-generated density for high, medium and low sulphur samples.

Figure 7 illustrates the top and bottom 10% of the samples based on equivalent gold grade. It is apparent that below 0.6 relative density, samples have significantly lower grade value. Also, there are two very distinct clusters of points that are distinguished by the rock type.

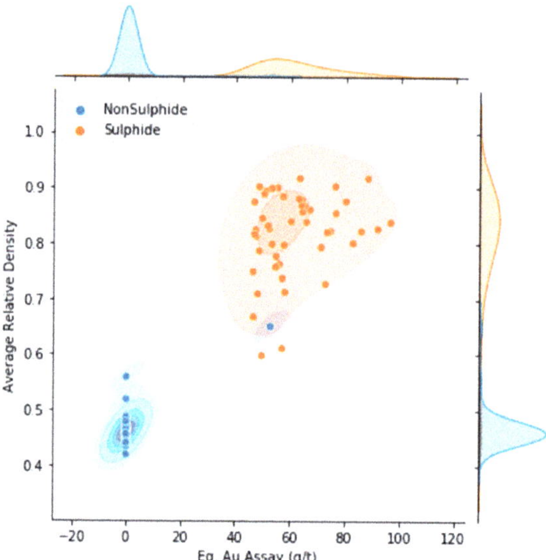

Figure 7. Scatter plot of equivalent gold assay versus average relative density.

Figure 8 plots all samples where similar characteristics can be observed, as illustrated in Figure 7. The distribution of the y-axis shown on the right side of the graph demonstrates the potential for removing more than 90% of the non-sulphide material below a 0.6 average relative density. However, when using the logarithmic scale to visualize the trends (Figure 8) when the relative density is below 0.6, there is no correlation between relative density and equivalent Au. This makes grade prediction difficult by using XRT. The tentative range of grades can be estimated from the scatter plot of equivalent gold assay average as shown in Figure 7, but this is not accurate enough for proper grades prediction.

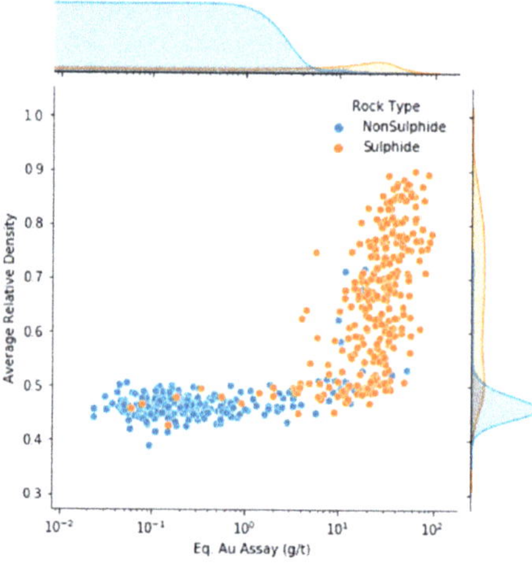

Figure 8. Scatter plot of equivalent gold assay versus average relative density-logarithmic scale.

3.2. Sortability Analysis

To assess the sortability, the average relative density value for each rock is put in order from the highest value to the lowest value. The metal recovery can then be calculated at a given mass yield. The mass is assumed to be equal for all particles in this analysis.

To plot a sortability curve derived from DE-XRT, each specimen is ranked by the lowest average relative density and the highest relative density. Then, the recovery curve is generated using the modelled metal grade of each rock specimen, as illustrated in Figure 8. The line graphs represent the cumulative mass yield and cumulative metal recovery (y-axis on the right side), while the bar graph represents the individual equivalent gold grade based on the descending order of average relative density for each particle.

As shown in Figure 9, 90% of sulphides can be recovered in approximately 55% of the mass. Furthermore, more than 95% of equivalent gold can be recovered in the same amount of mass. It is also important to note that at a 55% mass yield, the rate of calcium recovery almost doubles. This concludes that more carbonates are starting to be recovered when the 55% mass of rocks is being collected into the concentrate.

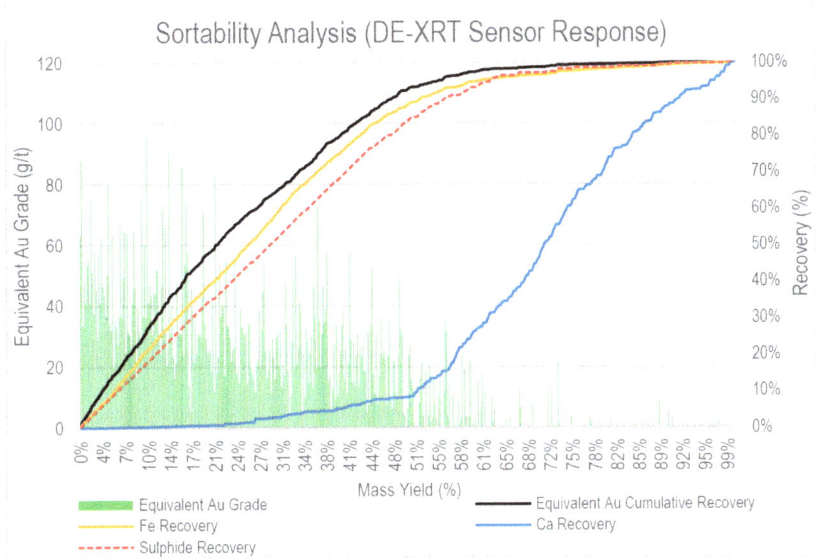

Figure 9. Scatter plot of equivalent gold assay versus average relative density.

4. Conclusions

In this study, the effectiveness of DE-XRT in characterizing material type for sensor-based sorting was explored. The new to mineral processing approach of a dual-material decomposition method was used to the DE-XRT image analysis.

For the studied ore, it was shown that DE-XRT sorting technology can recover 90% of the sulphides with a 55% mass yield. Although the DE-XRT was unable to accurately predict the grade of the rock based on the average relative density of the sample, it was able to separate sulphides from non-sulphides effectively. There is potential for applying regression methods for the sulphide rocks since there is a linear correlation between equivalent gold assay and average relative atomic density above 0.6, shown in Figure 7.

In this work, the dual-material decomposition method using linear components has been discussed, where the individual pixel's relative density is averaged for the entire sample. However, rather than averaging all the pixels within each sample, clustering methods can evaluate for nuggety mineralization. The clustering method can identify a cluster of pixels that share similar characteristics. For example, a quartz vein hosts a gold deposit where the amount of quartz vein in the sample can indicate the concentration of gold. An example of characterization criteria is that if more than fifty pixels of the nearest neighbour have a relative density value greater than 100, the sample can be characterized as an ore specimen. Furthermore, machine learning applications such as supervised learning can be used to train the classification of rock types based on a much larger set of samples.

In summary, the DE-XRT is an effective sensor that can be used to differentiate sulphides from waste. The dual-material decomposition method made it possible to use the characteristics of the sulphide materials and the non-sulphide materials to determine the composition of the sulphide.

Author Contributions: Conceptualization, Y.Z. and N.Y.; methodology, Y.Z.; software, Y.Z. and N.Y.; validation, Y.Z. and M.E.H.; formal analysis, Y.Z.; investigation, Y.Z.; resources, Y.Z.; data curation, Y.Z.; writing—original draft preparation, Y.Z.; writing—review and editing, Y.Z., N.Y. and M.E.H.; visualization, Y.Z. and N.Y.; supervision, M.E.H.; project administration, Y.Z.; funding acquisition, Y.Z. and M.E.H. All authors have read and agreed to the published version of the manuscript.

Funding: Sacré-Davey Engineering inc funded this research through the Mitacs Accelerate program.

Data Availability Statement: All data used in this article are from experiments conducted by the author. The name of the operations where the samples are collected are not disclosed.

Acknowledgments: From maintenance of the machine used for the experiments to lab supports, the Mining and Mineral Processing Engineering department has been very helpful throughout all experiments conducted for this paper.

Conflicts of Interest: The authors declare no conflict of interest.

References

1. Jones., M.J. *Tenth International Mineral Processing Congress, 1973: Proceedings of the Tenth International Mineral Processing Congress, Organized by the Institution of Mining and Metallurgy and Held in London in April, 1973*; Institution of Mining and Metallurgy (Great Britain): London, UK, 1974.
2. von Ketelhodt, L. Dual energy X-ray transmission sorting of coal. *J. S. Afr. Inst. Min. Metall.* **2010**, *110*, 371–378.
3. Tong, Y. Technical Amenability Study of Laboratory-Scale Sensor-Based ore Sorting on a Mississippi Valley Type Lead-Zinc ore (T). Ph.D. Thesis, University of British Columbia, Vancouver, BC, Canada, November 2012.
4. Bergmann, J.M. Latest developments and experiences in the beneficiation of coal using TOMRA X-ray transmission sorting machines. In Proceedings of the XIX International Coal Preparation Congress 2019, New Delhi, India, 13–15 November 2019; pp. 92–100.
5. Lessard, J.; de Bakker, J.; McHugh, L. Development of ore sorting and its impact on mineral processing economics. *Miner. Eng.* **2014**, *65*, 88–97. [CrossRef]
6. Zhang, G.; Chen, Z.; Zhang, L.; Cheng, J. Exact Reconstruction for Dual Energy Computed Tomography Using an H-L Curve Method. In Proceedings of the IEEE Nuclear Science Symposium Conference 2006, San Diego, CA, USA, 29 October–4 November 2006.
7. Kolacz, J. Advanced sorting technologies and its potential in mineral processing 2012. *AGH J. Min. Geoengin.* **2012**, *36*, 39–48.
8. Triche, B.L.; Nelson, J.T., Jr.; McGill, N.S.; Porter, K.K.; Sanyal, R.; Tessler, F.N.; McConathy, J.E.; Gauntt, D.M.; Yester, M.V.; Singh, S.P. Recognizing and Minimizing Artifacts at CT, MRI, US, and Molecular Imaging. *RadioGraphics* **2019**, *39*, 1017–1018. [CrossRef] [PubMed]
9. Rebuffel, V.; Dinten, J.M. Dual-Energy X-Ray Imaging: Benefits and Limits. In Proceedings of the ECNDT 2006, Berlin, Germany, 25–29 September 2016.

MDPI

St. Alban-Anlage 66

4052 Basel

Switzerland

Tel. +41 61 683 77 34

Fax +41 61 302 89 18

www.mdpi.com

Minerals Editorial Office

E-mail: minerals@mdpi.com

www.mdpi.com/journal/minerals